KU-625-095

Gregory Smithsimon

September 12

Community and Neighborhood
Recovery at Ground Zero

New York University Press • *New York and London*

974. 71044
SMI

NEW YORK UNIVERSITY PRESS
New York and London
www.nyupress.org

© 2011 by New York University
All rights reserved

Library of Congress Cataloging-in-Publication Data
Smithsimon, Gregory.
September 12 : community and neighborhood recovery at ground zero /
Gregory Smithsimon.
p.   cm.
Includes bibliographical references and index.
ISBN 978-0-8147-4084-2 (hardback) — ISBN 978-0-8147-4085-9 (pb) —
ISBN 978-0-8147-8671-0 (e-book)
1. September 11 Terrorist Attacks, 2001—Economic aspects—New York
(State)—New York. 2. Battery Park City (New York, N.Y.) 3. Buildings—
Repair and reconstruction—New York (State)—New York. 4. Manhattan
(New York, N.Y.)—Economic conditions.   I. Title.
HV6432.7.S65 2011
974.7'1044—dc23   2011020454

New York University Press books are printed on acid-free paper,
and their binding materials are chosen for strength and durability.
We strive to use environmentally responsible suppliers and materials
to the greatest extent possible in publishing our books.

Manufactured in the United States of America

c   10 9 8 7 6 5 4 3 2 1
p   10 9 8 7 6 5 4 3 2 1

Contents

Acknowledgments

For a project about taking space seriously, it makes sense to organize my acknowledgments spatially. Starting in Battery Park City, I want to thank the residents, local activists, and leaders whose welcome into their neighborhood made my research there possible. Alison and Robert Simko at the *Battery Park City Broadsheet* have been a considerable resource to me and to the community. Across Chambers Street from Battery Park City is the Municipal Archives. It's a pleasure to go there in no small part because of its director, Kenneth Cobb, who pleasantly initiates researchers into the wonders of the archive even though in that job there's no end to either wonders or researchers.

Riding the train north to Columbia University's campus, I owe thanks to many people: Sudhir Venkatesh, Herbert Gans, Charles Tilly, all of whom made this project better by persisting in asking me questions I didn't always want to answer. I was aided in my research on New Settlement Apartments by Alex Demshock while she was a student at Barnard College. Her connections with the area were of great use in gaining entrée to New Settlement staff. Lance Freeman, Susan Fainstein, and Lynne Sagalyn took time to consider my questions about Battery Park City and public spaces and gave me valuable direction in my research. I owe particular thanks to Priscilla Ferguson for lending me her office during the summer I was revising my manuscript. It was a perfect retreat where my most productive days were spent.

Those days would not have been quiet and productive had my mother-in-law Susan Simon not been nearby to watch my daughter. Susan also proofread an early draft of the entire manuscript, when it was a Herculean undertaking. I am deeply appreciative.

Brooklyn College's congeniality has been unsurpassable. The Department of Sociology has some of the most supportive and helpful colleagues anyone could ask for. Sharon Zukin has provided inspiration and insight all along the way. Tamara Mose Brown read drafts of every chapter. Alex Vitale's work provided a valuable foundation for my framing of the post-1975 crisis history of New York. Along with Brooklyn College, the Graduate Center at the City University of New York is home to an unmatched collection of urbanists, whose collegiality has always set them apart as much as their cutting-edge scholarship.

Further afield, geographically at least, my thanks to Louise Jezierski, who got me started down the path with an introductory urban studies course during my first semester as an undergraduate. She introduced the class to Jane Jacobs's *Death and Life of Great American Cities*, despite, I later found out, her misgivings about Jacobs's book for its lack of a good old-fashioned analysis of power. The book was what hooked me on urban studies, though I immediately shared Jezierski's sense that its story was incomplete. Part of the purpose of this project is to fill in the missing explanations about the role of power in shaping urban space.

Down in Maryland, I thank my parents, who have encouraged me without reservation in everything I did, despite the reservations they could reasonably have had. What I owe them goes far beyond this project, of course, but within the project, raising me in the suburbs—and taking us to Baltimore ethnic festivals, Washington, D.C., museums and even one memorable New York City Thanksgiving Day parade—no doubt primed my fascination with cities and public spaces.

At the end of this tour, I return home to Brooklyn to thank my family. My daughter Una was my first research assistant, accompanying me to the playgrounds of Battery Park City until they became part of her neighborhood. My son Eamon grew up looking forward to trips there as well. My wife Molly has been the enigma at the center of my urban research: a Manhattanite by birth with no particular affection for the city; someone who falls asleep reading most urban sociology but because of that is best able to tell me what parts of my project are most interesting; the person who was serious about making sure I had the time I needed to complete this project but just as serious about insisting that that time not go on forever. As with every other ambitious thing I've tried, I simply couldn't have done it without her.

September 12

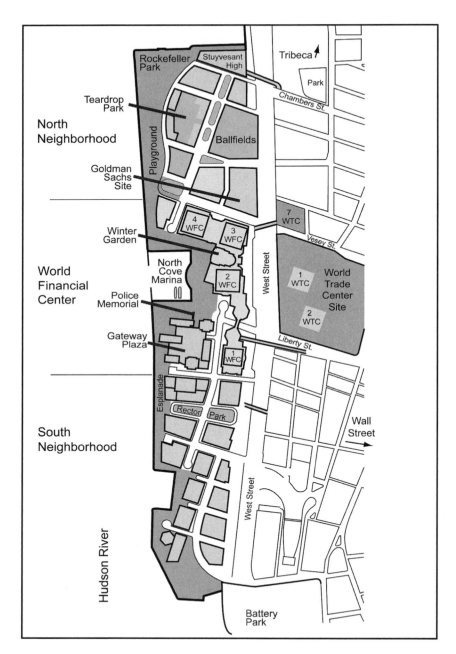

Map of Battery Park City

Introduction

LOWER MANHATTAN BROILED while Battery Park City was balmy. The contrast was evident in the tempo and temperature of the streets, sidewalks, and parks of the two adjacent neighborhoods. I pushed my daughter's stroller along the crowded sidewalks, making my way from Lower Manhattan to Battery Park City. It was a hot May day eight months after the Trade Center attacks of 2001. The narrow streets of downtown New York City clattered with the area's daily rhythms. Delivery trucks clogged Chambers Street, filling the air with dry exhaust. Heat reflected off the asphalt rutted by the constant passage of buses and taxis. Battery Park City was very different on days like this; a strong breeze blew off the Hudson River across the park and promenade and continued along largely traffic-free streets.

Though the two neighborhoods were intimately connected, their physical and social organization differed. Lower Manhattan, like much of the city, had long presented inequality at closer quarters, adjoining rich and poor, powerful and powerless. In the street-level mall of a bank office tower on Wall Street, bank employees hurried toward a subway entrance while heavily dressed homeless men took an air-conditioned respite at café tables. On the southern tip of Manhattan, American and foreign tourists picked their way through immigrant vendors selling New York City T-shirts, art, and souvenirs to wait under the sun for a ferry to the Statue of Liberty or Ellis Island. The heart of the Financial District, in front of the New York Stock Exchange, was preternaturally quiet behind roadblocks, security, and newly arranged concrete and steel barriers. Occasionally a trader in a color-coded suit jacket emerged from the Exchange for a cigarette. On Nassau Street, the most racially diverse street in the Financial District, workers from the surrounding financial firms, Pace University, and government buildings walked down the narrow pedestrian street, past mobile phone and jewelry stores to the lunch shops on the corners. Well below ground level at the vast World Trade Center site, construction equipment roared and beeped as it continued the slow process of clearing rubble from the site in anticipation of an ambitious redevelopment program. Each image was a reminder

The the north end of Battery Park City, one of New York's most literal approximations of the modernist "tower in the park" ideal.

that typical New York streets not only included a wide diversity of occupations, nationalities, races, and classes but also displayed substantial inequality in resources and privilege from block to block.

We rounded a corner, and the office towers and apartments of Battery Park City several blocks away came into view. The neighborhood was physically and socially organized on a different principle than the rest of Lower Manhattan. Rather than including the city's inequality in its tableau, it was nearly entirely affluent; the poverty that was dialectically attached to that prosperity had been shunted off to neighborhoods in Harlem and the Bronx. I found that the distinctive spatial organization of inequality that had produced Battery Park City shaped the community, its people,

Top: A stylish gate from the Downtown Alliance closes Nassau Street to cars. *Bottom*: Shops along Nassau cater to more middle- and working-class New Yorkers than the stores of the Financial District or Battery Park City.

In March 2002, reminders of September 11, sacred and banal, were ubiquitous in Lower Manhattan.

and beliefs. Place mattered not only in the provision of resources but in the construction of community and local politics.

Over the past forty years Battery Park City (given the name to distinguish it from, and associate it with, the venerable Battery Park just south of it) emerged as a comprehensive, state-planned development project of luxury apartments, financial sector offices, parks, stores, schools, and museums that was tucked to the side of the Financial District, stretching in the shadow of the Trade Center for a mile along Lower Manhattan's Hudson River waterfront. The neighborhood had drawn much attention, both positive and negative. It had enjoyed accolades as a rare urban development success story thanks to its supposed use of traditional, nineteenth-century urban street plans and architectural details. At the same time, it had been criticized in the literature of postmodern urbanism for being a "citadel"—a state-subsidized, exclusive neighborhood where the global corporate elite lived and worked, a luxury planned community.[1] Affordable housing was pushed to the outer edges of the city. To this controversial reputation had been added Battery Park City's experiences on September 11, 2001, when the World Trade Center towers, which were directly across West Street from both the office buildings and the largest apartments of Battery Park City, were struck by two planes. Right before residents' eyes, the towers burned ferociously for an hour and a half and then collapsed, killing nearly three thousand, temporarily choking out the sun and air for those who had not yet fled the neighborhood, and driving shards of the towers, their dust, and human remains through the windows of Battery Park City and into every crevice of residents' lives.

Like the ten-lane highway that separated Battery Park City from the rest of Manhattan, a wide swath divided opinions about the neighborhood's distinctiveness. Residents cited the area's visual beauty, its waterfront promenade and other pedestrian spaces, and its successful parks program as features they loved about the neighborhood. These aspects of Battery Park City's physical design, by drawing residents to congregate in public spaces, fostered public social interactions and the development of strong ties among community members. The neighborhood's isolation, both physical and socioeconomic, from the rest of New York, compounded by their shared experiences of September 11, further reinforced such ties. Yet as the urban planner Raymond Gastil accurately observed, Battery Park

A rollerblader on Battery Park City's Hudson River esplanade. The World Financial Center is in the background.

City, in part because it was a wealthy neighborhood built with state subsidies, had "generated more bile from the design community than any other project in New York."[2] Critics and defenders often used the same descriptor to embrace or deride this residential enclave of the Financial District: that it seemed like "a suburb."

This book is an ethnographic study of Battery Park City in the three years after the September 11 attacks, examining how residents used the rich social and physical infrastructure of the neighborhood as a basis to reestablish their community, and how in so doing they also reestablished the exclusive and privileged basis on which the community was founded. Though Battery Park City has been held up as an example of a "citadel" for a city's global elites, I find that this image suggests an earlier model for maintaining and defending an enclave of privilege from its immediate surroundings via barricades and surveillance. It is less prevalent today than what I call the "suburban strategy," which separates different social and racial groups by vast distances through programs of large-scale relocation and redevelopment.[3] And while earlier critiques identified the role of corporate and government elites in shaping exclusive enclaves, the interplay of the community and the spatial organization of this elite neighborhood creates an even more dedicated group of advocates for exclusivity: residents. I find that the exclusive space of Battery Park City interacts with the neighborhood's socioeconomic profile to influence residents' definition of the community, their positions on local issues, and their relations with people outside their community. This project contributes to the growing movement in sociology that recognizes space as a revealing product and producer of social relations and brings to critiques of the global city a detailed ethnographic account of the effects of exclusive urban spaces on community. The suburban strategy that Battery Park City represents is a paradigm of contemporary spatial organization in the service of segregation. Nonetheless, by understanding Battery Park City's spatial and social form we can see the potential to use urban space as the foundation of more inclusive and egalitarian social organization.

Entering Battery Park City

I eased the stroller down and up curbs. As my daughter and I headed toward Battery Park City, the neighborhood was still struggling to regain its footing, transformed into a staging ground for the recovery effort next door and cluttered with temporary power lines, tired rescue workers, improvised memorials, debris-filled barges, and still-boarded-up windows. But amid this disorder, residents were already trying to rebuild their neighborhood, physically and socially. Battery Park City's public spaces had long been widely criticized as historicist imitations of real urban parks and street life. But they would play a central role in the social processes that would unfold during the years of redevelopment planning. I had first researched Battery Park City two years before the Trade Center attacks to understand the workings of an upscale neighborhood with a reputation for seclusion, and to grasp how the design of exclusive public spaces shaped the social interactions that went on within them. I returned in the months after September 11 to conduct an ethnography of Battery Park City's public spaces, both to assess the original critique of Battery Park City as an ineffective imitation of real public space and a symbol of the stratified global city, and to observe how residents in this affluent neighborhood struggled to recreate a viable and hopefully vibrant community in the wake of considerable destruction. For the next three years I followed intense debate and planning by residents, community groups, local government, and real estate interests over the redevelopment of Battery Park City and Lower Manhattan. I interviewed private residents and community leaders, and attended meetings of the Community Board, local groups, and agencies related to the area's redevelopment. In the ensuing years I continued to follow up regularly. In observing the community, its political priorities, and its self-conception, I recognized a reciprocal relationship between the spaces and the people of Battery Park City as each reproduced elite social relationships by molding the other. The outcome of that spatial-social relationship was a distinctive neighborhood strong enough not only to recover after a devastating tragedy but also to serve as a model of spatial and social exclusion, catalyzing further development along the same lines and reshaping the entire city. To understand the interplay of public space and community, then, I was taking my daughter to the playground; Battery Park City has some of the best in New York City.

Before we reached Battery Park City, however, we had to walk past the World Trade Center site, which remained freighted with emotion. Construction workers and firefighters filed in and out. From across the street, I could see the staging area packed with stored equipment and construction trailers. Hundreds of visitors spread out along the construction fences to seek out views. What they saw was anticlimactic: everyone had come to see something that wasn't there. Still needing a souvenir of towers that no

Top: The popular waterfront playground in Rockefeller Park, in the northern half of Battery Park City. *Bottom*: "Pumphouse Playground," named for the World Trade Center water intake pumps beneath it, provided play space in between the World Financial Center (*in background, left*) and Gateway Plaza (*right*). It has since been renovated.

Break dancers practice in the park near Stuyvesant High School, the city's pre-eminent high school.

longer existed, visitors faced each other on the busy sidewalk and, at the moment a gap opened up in the stream of pedestrians pushing past them, snapped pictures of each other standing in front of open sky.

I could not pass the site easily. Whenever it came into view I felt a wash of emotions: sadness, which seemed maudlin on a sunny day, and embarrassment or guilt over that remorse. Beneath that, frustration, grief, and anger over what had happened. And always, always, a tone, beneath that, of reverence that New Yorkers instinctively felt toward the people who were—here the jingoists have the term right—*heroes* that day. Having read heartbreaking obituaries of parents who would never see their children again, and accounts of employees who had given their lives to help coworkers, I thought of them, too, but I always thought of the firefighters, whole squads of firefighters who had died. I don't know how long it was before I could see firefighters without tearing up, even in the most banal setting, as when they would pull a truck up and all get out, half in their bunker gear, to shop at my local grocery store. On the southern edge of the Trade Center site, along Liberty Street, not only were firefighters assisting in the recovery effort but more firefighters were standing outside the renovated Ten

House, the fire station that had been destroyed in the collapse and was now a stop on visitors' pilgrimage around the Trade Center. Walking past the site when I had somewhere else to go was a fraught process of trying to pass slowly enough to pay due respects, slowing down a bit more to gawk guiltily at the destruction or the progress of the cleanup, and hurrying up to avoid either seeming ghoulish or letting the emotion and tragedy of the place soak in so deep as to suck the wind out of me and demand that I detour from my trip to stop, take a deep breath, reflect on the events I had already reflected on too many times, and start walking again.[4]

To get to Battery Park City after exiting a subway stop not far from the Trade Center site, I walked down Liberty Street toward the rickety steel steps of one of the Trade Center's restored pedestrian bridges, which crossed the daunting highway called West Street. Battery Park City lay on the other side. The windows of the enclosed pedestrian bridge provided the clearest view of the Trade Center site, and visitors, tourists, local employees, and others stood looking out. From there we could see heavy construction equipment, indistinct far below us, removing beams, concrete, and debris. Yet even the void itself lacked visual drama equivalent to what had happened there: although, as press accounts often explained, the pit went down a full six stories below ground to bedrock, the site was so big—several city blocks in each direction—that it didn't look particularly deep, and one could only partially estimate the scale of each floor by seeing workers or vehicles all the way at the bottom or by counting the floors from the remnants of I-beam framework at the edges of the site. There was never any satisfaction to be had in visiting the site. From that view it was evident that cleanup was nearly complete, but the private and public developers overseeing the project expected that construction would continue for a decade or more.

Across the bridge in Battery Park City the pace was more relaxed, but few people took up the challenge of finding a way from Lower Manhattan across the broad lanes and fast-moving traffic of West Street. The few crosswalks on West Street were so long that the lights left people stranded halfway across the highway-sized road; others were stymied by the pedestrian bridges' elevators, which rose so slowly most people assumed they were broken.

The difficulty in reaching Battery Park City meant that it remained significantly apart from the rest of Lower Manhattan, even when tourists were crowded just across the street around the World Trade Center. Battery Park City's seven thousand luxury high-rise apartment units therefore sat in comparative serenity on the waterfront. The neighborhood's brick- and stone-clad rentals and condominiums were situated around exceptionally well-tended parks and public spaces. On entering, one immediately felt that Battery Park City was quieter, cooler, and cleaner than the rest of Lower Manhattan. The wind was still warm but made the air bearable in the tree-lined waterfront parks. Taxis dropped people off and trucks came and went, but less traffic made its way through the area, which had no streets that led anywhere else. Abundant parkland, trees, covered promenades, and air-conditioned mall walkways made passage through the neighborhood more leisurely than in the rest of Downtown.

Nearly every household was connected to New York's "FIRE" economy (Finance, Insurance, and Real Estate). The parks and promenades were not just for recreation but were part of people's daily commutes. In most households, at least one person walked to a high-paying job in the Financial District.[5] Ninety percent of men, and over 80 percent of women, worked in business, management, finance, the professions, or sales. Before the September 11, 2001 attacks, about one in five households made a quarter-million dollars a year or more. While the median household income was $41,994 in the United States and $38,293 in New York City, in Battery Park City it was $107,611.[6] (The *average* family income was over $192,000.) Over half of households benefited from investment income, compared to a third of households in the United States, while only 15 households out of 4,452 received public assistance, less than a tenth the percentage in the nation as a whole. The high wages were in large part accounted for by Battery Park City's proximity to Wall Street, and 20 percent of adult residents worked in the World Financial Center right in Battery Park City. Seventy-five percent of adults in Battery Park City had at least a college degree (compared to 25 percent of all Americans and 50 percent of all Manhattanites), and 42 percent had some form of postgraduate education. Business and professional degrees were particularly well represented, even in comparison to other wealthy neighborhoods nearby, like Tribeca.[7] A small number of moderate-income people lived in apartment buildings whose landlords took advantage of a mortgage tax credit program that required them to provide a small percentage of their apartments to moderate-income New Yorkers. But as the census figures bear out, the community was overwhelmingly one of well-educated, well-employed, wealthy households sustained by the financial industry.

Just as Americans often normalize their specific economic position, so many white Americans in particular normalize the segregation of the communities in which most Americans live.[8] And like most white communities, Battery Park City is racially exclusive. In 1999 New York City was 35 percent white, 27 percent Latino, 25 percent African American, and 10 percent Asian. Battery Park City's 7,951 residents were 75 percent white, 18 percent Asian, 5 percent Latino, and 3 percent African American.[9] (The places that Latinos and Asians in Battery Park City identified as their family's origins were largely the same places as those identified by the other Latino and Asian inhabitants of New York: China, India, Japan, and Korea for Asians, and Puerto Rico, Mexico, Cuba, the Dominican Republic, and Colombia for Latinos. However, countries with large working-class immigrant communities in the city, like the Dominican Republic and the Philippines, were comparatively underrepresented.) Considering racial exclusivity in terms of black residents, the fact that only three in every hundred residents were African Americans was not simply a result of Battery Park City being a wealthy community and African Americans being underrepresented among the wealthy. Citywide, 7 percent of households in the top income bracket (over $200,000) were African American, and nationwide 4 percent were. But in Battery Park City only 2 percent of households making over $200,000 were African American. In virtually every income bracket African Americans were underrepresented in Battery Park City, whether in comparison to the same bracket in New York City as a whole or to the same bracket in the nation (table I.1).[10]

Of course, the disruption of September 11, 2001 altered the population of Battery Park City substantially: several people who knew the neighborhood well estimated that half of all residents had moved out in the next two years, an estimate corroborated by surveys that found building vacancy rates of 25 to 75 percent in the year after the attacks.[11] The impact of September 11 was difficult to quantify further, however, because as a young, mobile population the community already had a relatively high turnover rate. The fact that only the decennial census provided data at the level of a single community stymied researchers' attempts to quantitatively compare populations before and immediately after September 11. On the basis of my qualitative research on Battery Park City, I found that the first new residents after September 11 did tend to be younger. Some were attracted, at least in part, by temporarily more affordable rents, but moving to Lower Manhattan at that time still required a serious commitment. (Some longer-term residents saw it as opportunistic that newcomers benefited from subsidies offered by the government.) But the neighborhood's demographic profile did not appear to be permanently altered by changes after September 11,

Table I.1

Tract	Median Household Income ($)	% Black	Total Population	% White	Location
9	105,456	5.9	1,111	74.6	Downtown (Exchange Pl.–South St.)
145	102,582	4.0	4,411	78.0	Columbus Circle (58th–62nd Sts. to 10th Ave.)
69	128,295	3.1	2,341	87.6	Greenwich Village (Houston–Christopher Sts.)
317.01	107,611	3.0	7,951	75.0	Battery Park City
21	128,384	2.8	2,407	78.1	Tribeca (Vesey–Reade Sts.), Broadway to West
167	93,335	2.3	6937	89.5	Upper West Side (78th–82nd Sts.)
149	96,588	2.3	5956	86.5	Upper West Side (62nd–66th Sts.)
153	94,583	2.0	9040	88.3	Upper West Side (66th–70th Sts.)
33	113,332	1.9	3,696	86.5	Tribeca (Hudson and Broadway)
120	115,430	1.9	3,965	92.8	Upper East Side (53rd–70th Sts.)
158.01	92,940	1.8	5804	90.3	Upper East Side (91st–96th Sts.)
86	110,330	1.7	7,267	83.2	United Nations (Tudor City, 33rd–53rd Sts.)
57	97,765	1.3	2535	88.9	Greenwich Village (4th–10th Sts.)
122	105,573	1.1	3,914	94.4	Upper East Side (63rd–70th Sts.)
114.01	108,107	0.9	1,484	94.5	Upper East Side (59th–63rd Sts.)
160.01	152,728	0.9	4172	94.5	Upper East Side (91st–96th Sts.)
140	97,430	0.8	7754	93.3	Upper East Side (77th–84th Sts.)
128	109,336	0.7	6,639	94.6	Upper East Side (70th–77th Sts., Park–3rd Ave.)
106.01	102,149	0.6	7,968	94.4	Upper East Side (54th–59th Sts., 1st Ave.–East River)
142	145,979	0.5	4980	94.8	Upper East Side (5th Ave.–Park, 77th–84th Sts.)
150.01	127,126	0.4	2,247	95.1	Upper East Side (5th Ave.–Park, 84th–86th Sts.)

African Americans are underrepresented in Battery Park City by any measure. Even so, compared to the twenty Manhattan census tracts with the most similar median household incomes, Battery Park City ranks near the top for percent black and has virtually the smallest percentage of whites. *Source*: 2000 Census.

2001. When rental subsidies expired two years later, rents remained high, and the neighborhood continued to be extremely expensive.[12]

Entering Battery Park City meant entering a world very different from the rest of Manhattan. Built on landfill that had covered over the docks along this stretch of the Hudson, Battery Park City physically embodied the ascendance of the financial industry over New York's history as an economically diverse, working-class port city.[13]

Battery Park City also reflected how the financial industry had gained dominance not by private market power alone but by public financial assistance. It had been conceived in the early 1960s by banking elites who, in hopes of shoring up their position in the Financial District, had convinced government officials to finance the planning and initial construction of a project that would become a home for both major corporations and the managers and executives who worked in them. Observers had noted that

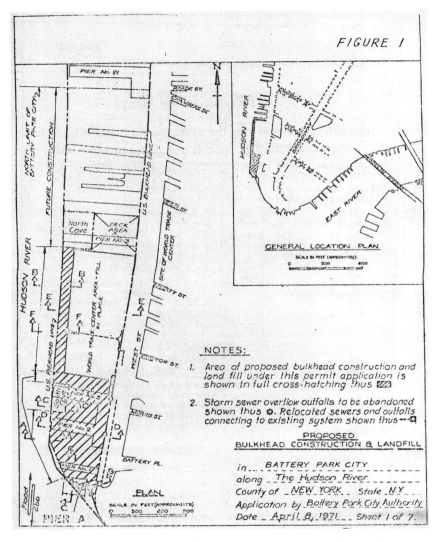

The Financial City buries the Industrial City: 1971 plans by the Army Corps of Engineers show construction of part of Battery Park City from landfill from the World Trade Center. North of the landfill, new land is created by the building of a deck over the Hudson River PATH train tunnels. North of that, outlines of the existing piers are shown. The size of the neighborhood was determined by the length of the piers, as shown by the border of the U.S. Pierhead line. (Army Corps of Engineers, *Battery Park City Authority. Hudson River, New York. November 15, 1971* [Springfield, VA: National Technical Information Service, 1971])

it was something previously thought to be an oxymoron: welfare for the wealthy, a state-subsidized luxury housing project. To maintain this privileged situation, the state first financed the project, then structured local taxation to exempt residents from regular property taxes but provide lavish funding to maintain the neighborhood. New York State retained tight control over the neighborhood by establishing for Battery Park City its own separate zoning regulations and tax revenues and by retaining ownership of the land, carefully vetting the developers it hired to build there, and leasing lots gradually enough (over thirty years) to allow only buildings constructed for New York's comparatively small luxury market. Since the neighborhood's inception, critics had been outraged at this oasis of wealth, state largesse, and unique private privilege and public oversight that ignored pressing needs elsewhere in the city, and they were frustrated that state officials bestowed on the already-wealthy Battery Park City privileges that no other neighborhood enjoyed.

Since Battery Park City's creation, scholars who have critiqued "global" cities like New York, London, or Los Angeles as strategic locations for the global concentration of wealth and power created under contemporary capitalism have pointed to the neighborhood as the preeminent example of a "citadel." The term evokes medieval Europe, where a citadel was a fortified stronghold that could command the surrounding city, and today refers to a physically and symbolically defended enclave in which managers of global capitalism—those that oversee its "command and control" functions—live, work, and play. As a redoubt for corporate power and a barrier to democratic interaction, citadels like Battery Park City are often rhetorically and literally at the center of critiques of global capitalist cities.

The classification of Battery Park City as a citadel draws on the "citadel and ghetto" typology of John Friedmann and Goetz Wolff's pathbreaking writing on "world cities." Elements of their description of a citadel are reflected clearly in Battery Park City: "With its towers of steel and glass and its fanciful shopping malls, the citadel is the city's most vulnerable symbol. Its smooth surfaces suggest the sleek impersonality of money power. Its interior spaces are ample, elegant, and plush. In appropriately secluded spaces, the transnational elites have built their residences and playgrounds."[14] The citadel is a concentrated enclave defended by private security forces, physical barriers, and economic and racial exclusion. The ghetto presses against its borders. A product of the "inequality and class domination" of globalization, citadels are designed to host the functions, residences, and recreation of the class that manages the global economy. This dialectic of a bifurcated city has become a recurring theme in writing about global cities. Both Saskia Sassen and Manuel Castells, for instance, see elites' creation of segregated residential areas as a defining element of the new global, or informational, city. Mike Davis has concluded that citadel elites wage class warfare against ghetto residents primarily through their physical restructuring of the contemporary city.[15] Sharon Zukin has argued that property rights, rents, zoning laws, transport systems, and symbolic systems of control are used to create central locations in cities that are themselves spatial reflections of the market economy.[16]

The primary means of entering this citadel have been the pedestrian bridges over West Street. Initially they connected Battery Park City directly to the World Trade Center. In the years since September 11, pedestrians have unceremoniously entered one reconfigured bridge (eventually, two new bridges) from the streets surrounding the Trade Center construction site. The pedestrian bridge that my daughter and I crossed on that day in 2002 led to the World Financial Center, which occupied the middle of Battery Park City. The World Financial Center was the economic core, even if not the social center, of the residential community that flanked it to the north and south. Before September 11, the managers of global capitalism had worked inside its four connected towers (plus the New York Mercantile Exchange building next door). Some of the buildings were occupied again, and all would eventually reopen. The World Financial Center, a contradictory mix of granite opulence and shopping mall–style normality, was one of the "command-and-control centers" described almost mythically in critiques of the global city. People whom the author Tom Wolfe famously dubbed the "Masters of the Universe" worked here in the large offices of companies like Merrill Lynch, American Express, and (before its spectacular bankruptcy) Lehman Brothers.[17]

I had conducted field observations in the public spaces of the mall and its outdoor space in 1999, when the Trade Center still towered over Battery Park City. At the time, the towers were connected by the Winter Garden Mall, a luxurious if underused shopping mall whose major function was to provide a range of themed lunch options to the office workers upstairs. The west sides of the buildings faced the North Cove Marina, and the open-air seating of restaurants on the ground floor gave the financial workers a view of the water at lunch and dinner. Beyond the restaurant tables were chess sets on outdoor tables to allow lunchtime games. In the mix of office workers, men outnumbered women. People sauntering down the waterfront promenade and through the pristinely maintained parks set a breezier tempo here than in the rest of Lower Manhattan. Their conversations were thick with references to financing, planned credit card promotions, "financial officers," and "IPOs." "*Concierge* just means 'janitor' in French," one well-dressed man explained to another.

The World Financial Center exemplified the contradictory nature of Battery Park City in the years after September 11: the epitome of corporate power, in 2002 much of it remained seriously damaged and unusable. One manager at a consulting firm that had employed three thousand people at the World Financial Center before September 11 described the scene when he returned to oversee a small group of employees from the firm who salvaged documents, materials, and employees' personal effects from the office a week after September 11. "It looked like a war zone. Unbelievable. There were no windows. The destruction that was in the space—it was very, very sad." In the beginning they worked in the dark. "There was no electricity. No running water . . . Some of the things were on the thirty-seventh floor, and no elevators. You really didn't even have security."[18] He remembered seeing the bodies of firefighters pulled from the rubble of the Winter Garden. Working in the shell of the building was difficult. "For the first six months I came here, I kept feeling like I was coming back to a different location, and eventually I'd be able to come back to my location where the Trade Center was at and where my buildings were at . . . Everything was so turned upside down." In a place that was (before and after September 11) synonymous with luxurious corporate power, employees, construction workers, and rescue workers encountered a landscape of destruction during the months and years of recovery work.

From a part of the World Financial Center that was already reopened, I walked to the adjoining North Cove Marina. The docks in the marina had been a showplace for the largest yachts in New York Harbor, luxury boats registered in the Caribbean that prominently carried small submarines, mock helicopters, and other vehicles on their decks. Those ships were no longer in the marina, a result of both Battery Park City's widespread disruption and the marina's leasing difficulties. But some large sailboats, as well as smaller ones used for sailing classes, had refilled the North Cove. The portable volleyball court, damaged by heavy equipment during rescue and recovery, was once again being used by nearby office workers. At the south edge of the marina, a temporary memorial to fire, police, and other emergency service workers attracted a small but steady audience to a white tent under which people pinned pictures, notes, and mementos. To the east side of the marina, joining two towers of the World Financial Center, was

A restaurant worker talks with his family during a break (*right*), while people lace up rollerblades to skate on the promenade (*left*). By January 2002, several businesses had reopened, and some activity had returned to the neighborhood. But the neighborhood was emptier, and more intimate, than before.

the arched glass-roofed enclosure of the Winter Garden Mall. By the first anniversary of September 11 it would reopen and once again be a tightly controlled, palm tree–shaded oasis whose glass walls and security guards kept out bad weather and homeless people while welcoming shoppers, office workers, and mothers and babysitters looking for a place to bring the infants they cared for. But in the spring of 2002 it remained closed after the interior was partially buried by the collapsing North Tower of the Trade Center.

Though the mall was closed, other neighborhood stores had already reopened. The grocery store, Chinese restaurant, nail salon, and others were back in business. And the "multiethnic managers of the global economy" I had observed in 1999 were eating lunch once again at the open tables of restaurants around the marina, their conversations once again focused on financial products and corporate strategies. By day, these office workers in open-necked dress shirts and khaki pants seemed harmless enough. But one night during my research I saw a dozen of them led out in handcuffs for currency fraud.[19]

Top: World Financial Center employees fill the North Cove at lunchtime, September 2003. *Bottom*: Two months later (November 2003), an agent in an FBI jacket walks in front of a van carrying employees of a firm at 2 World Financial Center, arrested for rigging currency trading. Television camera crews film their departure.

Frequent visits to such a site can lend a sense of the "ordinary" to physical concentrations of economic power that are not ordinary at all. Hannah Arendt has commented on the banality of the face of evil, and for me, an avowed anticapitalist (though one who had paid the bills with more than a few temp assignments in offices much like these), a location like the World Financial Center was simultaneously damningly opulent and disarmingly mundane.

Exclusion by Design

There have been few in-depth studies of Battery Park City. The most provocative claims about its citadel effects are found in works that hold it up as representative of a larger trend in cities or capitalism.[20] First, critics have linked Battery Park City's physical exclusion to its economic exclusivity. The novelist and commentator Phillip Lopate, for instance, has related the neighborhood's elite status to the exclusionary effects of its spatial organization: "How public *is* public space, when it has been embedded in a context that raises such formidable social barriers that the masses of ordinary working people (not to mention those out of work) would feel uncomfortable entering it? How many poor families may be expected to cross the raised bridge into that citadel of wealth, the World Financial Center, and wander through the privileged enclaves of South End Avenue and Rector Place before reaching their permitted perch along the waterfront?"[21] Lopate found that Battery Park City excluded poor families in two concurrent ways: by pricing them out of the neighborhood and by encircling it with a moat that one could cross only by challenging the ramparts of wealth and privilege. Economic and physical exclusion went hand in hand.

Other critics take this argument further by linking the class exclusivity and spatial exclusivity to a larger, intentional restructuring of space in the service of an increasingly harsh global capitalism. In M. Christine Boyer's account, for example,

> The traveler is blocked from the river by streams of roaring traffic on West Street or arrested in the subterranean maze of the massive platform that serves as the base for the twin towers of the World Trade Center. Just beyond reach, jutting out from the Hudson River waterfront, lies the new development called Battery Park City. The pedestrian is really not welcomed to this new public space, for she or he has to painstakingly and cautiously

cross lanes of highway traffic with no obvious point of entry in sight. . . . Battery Park City stands as yet another isolated city tableau in the contemporary game of spatial restructuring.[22]

Francis Russell similarly concludes that the physical barriers to Battery Park City were designed in the master plan to create a protected, elite neighborhood. "The physical isolation of Battery Park City from the rest of Lower Manhattan, its uniformity of appeal, and its economic exclusivity all work to remove its occupants from any meaningful contact with the adjacent authentic urban experiences. *This can only be seen as intended* and expedited by the development controls."[23] For Russell, Battery Park City's carefully designed exclusivity was too artificial for the neighborhood to be an organic, authentic part of the city.

Russell, like many other critics of Battery Park City, concludes that the neighborhood's high rents, its physical isolation from the city, and its residents' presumed social isolation must all equally reflect developers' exclusionary preferences. Similarly, for Boyer, elites both created a wealthy community and isolated it for the same purpose: to restructure the city to serve the demands of postmodern capitalism.

Battery Park City, in these views, is much more than a wealthy neighborhood. Because it was built to house the wealthy, it was made into a filtered space that would allow access only for the privileged, a space carefully designed as a defensible garrison from which to direct global capital, but one that revealed the ruling class as ultimately too insulated to engage in the real life of the city. Sympathetic as I was to these views when I first entered Battery Park City, I soon found that while elements of exclusivity were real, the community was far more engaged, complex, and promising than any of these critiques allowed.

In recent years, urban sociologists have studied the spaces in which social processes play out to better understand social phenomena.[24] But the most influential research on space has sprung from architectural critiques presuming that physical space reflects social relations in clear, legible ways: physically exclusive spaces produce social exclusion, whereas open designs create a more democratic polity. In this view, a building's facade alone can legibly communicate the intent of its owners or the social conditions of its creation.[25] Likewise, a physical barrier like West Street would be evidence of elites' desire to establish a barrier. But the city is not a text. If it were, the building facades alone would contain enough information to reveal the purpose of the built environment. Instead, understanding the meaning of the built environment requires an examination that combines the text with context, the facade with the actual history of its design and construction. Then the relationship between space and the people who inhabit it becomes evident.

Continuing our walk, my daughter and I moved north from the office towers and shopping mall of the center of Battery Park City. Residential neighborhoods of high-rises occupied the spaces both north and south of the World Financial Center. The apartments were built to design standards so exacting that a Battery Park City official had to approve the color of each order of bricks. Those standards ensured that, to someone walking along the quiet streets, the buildings would look much like the late nineteenth-century luxury apartment buildings of the Upper West and Upper East Side neighborhoods of Manhattan. The newest building guidelines established standards for environmental sustainability, so in the North neighborhood we could see the city's first "green" apartment building under construction. Called the Solaire, it was a luxury project due to include not just solar panels on the sides but bamboo floors, low-odor paint, a "green roof" with gardens to catch rainwater, and a second set of water pipes to use recycled water in toilets. Just across the street from the construction site for the green building, in Governor Nelson Rockefeller Park, which ran along the edge of the river, was the largest and most popular of the neighborhood's three playgrounds.

The playground put on public view both the long-standing stratification in Battery Park City and new anxieties. My daughter and I entered through the steel gates. It was crowded, as it was most days. Nannies, mostly black women who had emigrated from "the Islands" of the Caribbean, sat in the shade of a trellised seating area at one end, feeding infants in eight-hundred-dollar strollers, or stood near the sandbox playing with their charges.[26] A white college-aged woman hired to care for children, her blonde hair pulled neatly back, interacted little with other nannies or with parents, though she was often mistaken for a parent herself. Stylish and casually dressed mothers and a few fathers moved around the playground, playing with their kids and talking with neighbors. I regularly brought my own daughter here so that she could play and I could catch up with people I knew in the neighborhood. I had recently joined a larger research project of the Russell Sage Foundation on the effects of September 11 by submitting a grant proposal for what I hoped would be an effective example of research multitasking: a study of Battery Park City's public space that would begin by my taking my one-and-a-half-year-old daughter to the playground and simultaneously gaining entrée with parents and other residents. I hoped the project would allow me to do field research even though I didn't have child care. Experience elsewhere had already taught me that parents were much more comfortable starting up conversations with each other than with other strangers in a park.

As it turned out, the playground was revealing, even though in the spring of 2002 many residents did not socialize as easily as I had expected. The mood was lighthearted, but the shadow of September 11 could flit faintly across the sky at unexpected moments. I exchanged greetings with a young mother who had attended a recent Community Board meeting. She worried about the health risks of the exhaust from the ferries that docked within sight of the playground. Two other mothers stood by the sandbox discussing their workout routines, which included personal trainers and combinations of different styles of yoga, then went on to talk about their children's food allergies. Just as they neared the subject of asthma and environmental allergies, there was a break in the conversation. There was plenty in the busy park to distract them. But there were also worries about ongoing contamination of apartments in the neighborhood from the toxins and heavy metals blown into homes on September 11, and those parents who had already decided it was safe to move their families back into Battery Park City generally steered their conversations away from that topic. Shadows of September 11 appeared again when a gray, low-flying military helicopter zoomed along the same Hudson River flight path taken by the planes that morning the previous September, and parents turned their heads to see what was causing the noise, attempting the casualness of people who don't want to be caught looking at their watches to see how late it has gotten. A four-year-old playing nearby began talking to my daughter, and then to me. "I used to live in Battery Park City," he informed me, "but now we live on Abbey Street." Many families who had moved away after September 11, either because their apartments were uninhabitable for months or because they no longer felt safe living in Battery Park City, returned to the playground to meet up with friends from their former neighborhood, comparing life in the city to the suburbs, or praising their new Brooklyn neighborhood. Socializing in the playground was hardly child's play. Parents had to navigate their conversations around topics like environmental risks that might offend other parents, avoid commenting on anxiety-provoking aircraft so as not to sound overly anxious, or explain their move out of town without sounding defensive or critical of others who had stayed. The children had fun, but parents had to be cautious.

Evidence of Battery Park City's affluence showed up not only in its landscapes but also in its special resources for children. The park house just past the playground was so well stocked it would be the envy of any other park in the city. Staff invited children to play table hockey or borrow balls and toys. On the grass, toddlers pushed and climbed foam-rubber blocks as big as themselves. A rack offered parents picture books to read to their children. The ones in Spanish went largely unused. Beyond the park house, students from nearby Stuyvesant High School, the city's preeminent magnet school, lounged in small groups and played in the handball courts.

After lunch in the playground, we packed our things back in the stroller and headed home, taking a long walk down the waterfront promenade that ran the length of Battery Park City along the Hudson River, to see what was happening in the always-active neighborhood, and to observe the latest physical changes to a place that would still be in the process of restoration more than a year after September 11.

As I left the neighborhood at the end of my visit, the rest of the city felt simultaneously more difficult and more welcoming. I always enjoyed visiting Battery Park City but liked it most when I brought friends there. There was a particular satisfaction in introducing the neighborhood to newcomers who didn't know about the riches it concealed. At the end of most visits, it was hard to leave the soothing calm of landscaped gardens and tree-shaded walkways. Trading the sounds of sailboats clinking in the marina for a subway car banging into the nearest station was a jarring transition. But the train and its passengers were a welcome connection to a much larger city.

Overview

Several themes evolve over the course of this book. At the most general level, this project examines how elites initially shape space and how people and place then reciprocally affect each other. The suburban strategy ultimately employed in Battery Park City fostered spatial segregation and a vibrant exclusivity. These features set the stage for the neighborhood's approach to recovery after the devastation of September 11, when residents made intensive use of the neighborhood's public spaces and organized to reproduce the spatial exclusion that shaped their community. Each chapter examines a cycle of the reciprocal relationship between Battery Park City's people and the space that they collectively occupy, and considers the social implications of suburban space in the global city.

Chapter 1 examines the urban planning history that led to the creation of Battery Park City to show how the exclusive plans that elites endorsed for urban development have evolved over the past fifty years. The sequence of plans reflects the changing social context in which they were designed, and thus indicates how physical design, when studied alongside debates from that period and the broader historical context, can illustrate the vision of the city held at a particular point in history by the coalition of prodevelopment interests that the urban scholars John Logan and Harvey Molotch call the "growth machine." In particular, the history of plans for Battery Park City reveals four consecutive stages in elites' attitudes toward New York that are reflected in New York's public spaces and elites' projects.[27] The earliest plans reflect Battery Park City's working-class origins, but that history was soon eclipsed by plans to *privatize*, then *filter*, and finally *suburbanize* the neighborhood's design. Each stage reflects elites' shifting views of the city: first as a place of disorder from which they needed to defend themselves, then as a site that could be selectively gentrified, then as a city that could be recolonized on a large scale. At each turn Battery Park City, its plans, and its public spaces give early indication of these evolving approaches to elite urban planning.

Battery Park City's planning history also explains how the forces of global capitalism created particular physical forms in global cities. Just as important, it lays out the historical contingencies that shaped the contemporary neighborhood. Strangely, Battery Park City long had a reputation for lacking history. In 1989, journalist Richard Shepard noted urbanists' complaint that Battery Park City was a simulated neighborhood because (in contrast, for instance, to the storied past of monument-rich Battery Park) it lacked any real history of its own. Shepard sought to present the blank slate that so troubled postmodern critics as an asset. "What's exciting about this landfill is that history is just beginning."[28] In fact, as chapter 1 demonstrates, by the time ground was broken for the first building, Battery Park City already had a long and contentious history.[29]

The outcome of these plans—a demographically exclusive place with distinct but less extreme spatial exclusivity—accounts for what I describe as "vibrant exclusivity." Battery Park City demonstrates the importance of histories of social conflict in shaping space and provides a framework for understanding how the design of a neighborhood's space shapes the community that occupies it.

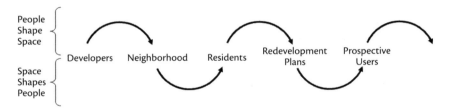

Space and society are interlocked in a reciprocal relationship, as people seek to shape space to reproduce their social privilege.

The history of elite plans for Battery Park City that I examine in chapter 1 also reveals the first stage of a *reciprocal relationship* between space and social relations that runs through the book. The reciprocal relationship is a cyclic one in which people shape space, then space shapes people, and then people shape space again, so that space is constituted by social actors seeking to reproduce their own particular social position, and the space then reproduces and constitutes social relationships that take place within it, setting the stage for a new round of social actors to attempt to mold space once more to reflect their social position. Thus, in this first stage of the reciprocal relationship, elite Downtown actors sought to shape the area that would become Battery Park City into a space that would reinforce the privilege and objectives of the executives in the Downtown-based financial industry. Actors were intuitively aware of this relationship and therefore hoped that their shaping space would allow them to reproduce the particular relationships of social privilege of which they were beneficiaries.

Chapter 2 continues the examination of the suburban strategy that shaped Battery Park City, using ethnography's extended-place method both to identify the major privileges bestowed on the neighborhood and to assess the contributions Battery Park City has returned to the broader city. An examination of New Settlement Apartments, Battery Park City's "twin" project in a low-income neighborhood in the Bronx, presents the shortcomings of assumptions that state-sponsored luxury projects can adequately contribute to the larger community. While the financial advantages Battery Park City has enjoyed have been real and substantial (including its ability to get city services without paying taxes and to put its property assessments into local projects rather than city coffers), the much-touted public benefits of Battery Park City for the rest of the city (particularly the plan to use surplus Battery Park City revenue to fund affordable housing elsewhere) have been largely illusory and have even reinforced racial and economic segregation. The special privileges that accrue to Battery Park City residents are significant because subsequent chapters show how these privileges interacted with the neighborhood's spatial organization in such a way that residents defined their community through exclusion.

Chapter 3 looks at residents' opinions about their neighborhood to document the ways in which they have developed an exclusive attitude toward the community, outsiders, and the relationship between Battery Park City and the city beyond. This attitude turns out to be the product of both the privileges of the neighborhood and the spatial organization of the place, so that programmatic elements and design elements contribute to the socially shared conception residents have of their community. In particular, the protection of privilege and definition of the neighborhood by physical boundaries leads to a spatial definition of community, a particular way of defining community that reflects both space and social position.

While earlier critiques of the global citadel have blamed the machinations of global forces and anonymous planners for constructing an enclave for the global elite, this chapter demonstrates how residents become highly invested in reproducing their citadel's exclusivity. Residents, while working hard to create their community, not only explicitly defined their neighborhood as secluded and exclusive but played an important and previously unrecognized role in preserving Battery Park City's isolation. In this way, the reciprocal relationship has continued as residents are shaped by the isolation of their community and define their community in exclusive terms.

The collapse of the World Trade Center towers showered debris onto Battery Park City and forced the evacuation of the neighborhood. Chapter 4 examines how residents recovered from the devastation of their social community and their physical surroundings.[30] More than any other community, Battery Park City residents experienced the attacks of September 11 as a prolonged dislocation of their normal lives. The collapse of the towers literally cut them off from the rest of the world. While this made life in Battery Park City practically and psychically more difficult, it demonstrated how the unintentional creation of isolated, "community-only" space actually facilitated residents' recovery, for they relied on the shared spaces as places where they could socialize, talk over the trauma with others who had experienced it, and rebuild the community networks that had been disrupted by the attacks. In this way, community space, though less accessible and less public, proved immensely valuable to the people who lived there. The contested nature of public space that a community claims as its own remains evident, however: the sense of a community united against outside threats simultaneously set the stage for conflicts between residents and the rest of the city as recovery and redevelopment plans progressed.

Chapter 5 examines some of those conflicts within Battery Park City as residents approached the one-year anniversary of September 11. Residents were concerned about the way it would bring outsiders—whom they called "tourists"—into their enclave. Their opposition to a public role for Battery Park City after September 11 was likewise on display in their opposition to a series of memorials in Battery Park City. In both cases, residents' anxieties about changes to the neighborhood and its use by a larger public were experienced and expressed spatially.

In chapter 6, the reciprocal relationship comes full circle: people shaped the space, then the space shaped residents, and now residents work to shape the space anew. Initially, elites conceived of the neighborhood as an exclusive citadel. Then the neighborhood's spaces contributed an exclusive perspective to residents' definition of their community. Finally, during debates over Trade Center redevelopment, residents sought to reproduce that spatial exclusivity by endorsing redevelopment plans that would ensure another generation of physically exclusive designs. After state agencies unveiled plans that would have made the thoroughfare that separated the neighborhood from the rest of the city less of a barrier, residents banded together to oppose the changes. Soon after, they lobbied to build a bus parking garage on top of the footprints of the Twin Towers in order to keep visitors and their buses out of Battery Park City. Ultimately, residents sought not to eliminate nuisances like the parking garage and the highway but to strategically deploy them, right in their backyard, to achieve exclusive community objectives. This approach to space reflected not a NIMBY (Not in My Back Yard) approach to nuisances, but DIMBY (Definitely in My Back Yard), in which residents mobilized in favor of disruptive urban elements because (as described by the reciprocal relationship) those structures could be used to concretely reproduce existing exclusive social relationships. The positions residents took on both issues can be understood only with an appreciation of the social contribution that space made to the community, demonstrating the need to consider the role of space in the study of community conflicts.

Battery Park City is a vibrant, cohesive, yet exclusive community of residents who enjoy social and municipal benefits that all New Yorkers deserve but many are denied. The neighborhood's totality—its popular parks, its orphaned affordable housing in the Bronx, its residents' need to be separated from other neighborhoods—provides a new and more accurate portrait of the shape contemporary cities are taking and proposes spatial perspectives for studying community conflicts. While traditionally citadels have been described as isolated outposts of opulence surrounded by hostility and deprivation, today adjoining citadels form larger, contiguous, affluent areas. Contemporary elite neighborhoods fit into a new urban model that replaces block-by-block diversity with the suburban strategy of expansive, homogeneous, unequal developments. The impact—on urban culture, on citizens' willingness to support public expenditures, on local and regional politics—is already proving to be substantial. Battery Park City is less an example of what is to come than a powerful part of what already is.

Studying Up

Ethnography has long been used in urban sociology to develop in-depth insight into the social organization of communities, the wrenching effects of community disruptions, and the process of constructing shared meanings and culture. I had the good fortune to train with exceptional ethnographers, and when I began my fieldwork ethnography seemed like a natural way to approach the subject. But in using ethnography to contribute to the literature on the exclusive design of global citadels like Battery Park City, I applied ethnography in two ways it is not typically used.

First, this is a public space ethnography. Adding to the many ethnographies that study intimate domestic spaces, workplaces, and other private areas is a tradition of "street corner" ethnographies that study social interactions in public settings.[31] My project, however, specifically sought to study not just the social interactions in a place but the physical place itself. Like Mitchell Duneier's *Sidewalk*, it examined not only what went on in the space but how certain social spaces allowed certain kinds of social relationships to develop and be sustained.[32] For this reason, my observations were not only of the people but of the places and, most importantly, the interaction between the public spaces and the social interactions that occurred in them and reshaped them.

Second, ethnographies are rarely of elite communities.[33] In urban sociology, ethnography is most often used to present a socially richer, more fully formed view of disadvantaged communities of people, including African Americans, immigrant and ethnic groups, and poor and working-class communities—all people whose lives and stories are given short shrift or mercilessly distorted in political rhetoric, casual conversations, and public policy discussions. In addition to researchers' other specific goals, ethnographies have often been successful in presenting the lives of people who are often disrespected, stereotyped, or ignored, and in chronicling the ways that social inequalities imposed by more powerful groups affect people who bear the brunt of such inequalities.

In contrast, this study was an *elite ethnography*, an effort to look up the social ladder, not down. Battery Park City residents' incomes rank the area the twenty-first highest out of New York City's 2,200 census tracts, sharing the top 1 percent with neighborhoods such as the Upper East Side and Upper West Side, Greenwich Village, and Tudor City near the United Nations. Battery Park City's status posed several challenges beyond the fact that the high cost of housing prohibited me from living in the neighborhood, as ethnographers often do for ethnographic community studies.[34] It is, frankly, difficult to get to know people well and remain publicly critical of them, even when they deserve honest criticism, and I found myself naturally cautious in leveling criticism at people whom I was close to and who were socially superior to me in age, rank, wealth, and status. Maintaining an appropriately critical or analytic stance while conducting an elite ethnography is inherently challenging.

Balancing a critical perspective while remaining true to residents was challenging for other reasons as well. Like most people, I have never lived in a wealthy neighborhood. While the wealth and privilege of Battery Park City were not normal to me, they were not utterly unfamiliar. I had attended Ivy League universities for college and graduate school, felt as though I had assimilated well, and had made many upper-middle-class friends there. Thus when I was in a place like Battery Park City I imagined that I visually fit in with the residents and that I could "pass" as one of them. Unlike most of the nannies, for instance, I was white, as were most residents, and my education put me in the upper socioeconomic quintile even if my income, wealth, and occupation did not. Further, I assumed that plenty of Battery Park City residents had not come originally from wealthy families, so our social distance was further reduced. This bifurcated sense of belonging and not belonging would continue throughout my research. Though I felt I fit in for the reasons I have described, I was not a resident, in a community where residence was the central marker of belonging. I was generally younger than other parents there, and I had different interests and career goals than most residents. I came to know many residents quite well, to like them a great deal, and to care about them and what they felt. But we were not so close that we socialized, with some important exceptions, outside the context of my research. I spent time in people's homes, but to interview them, not to participate in their lives. (Mine was, after all, a public space ethnography, not a study of their domestic lives.)

At the beginning of the project, it also felt uncomfortably opportunistic to study Battery Park City. I had wanted to study public space in New York, and from earlier projects I recognized that people articulate their views about public space only when there are plans afoot to alter it. During the economic downturn after September 11 it gradually became apparent that the only public spaces that would be under heavily contested redevelopment would be those surrounding the World Trade Center, where, along with Battery Park City, I had already conducted preliminary fieldwork. So many researchers had swooped down on the neighborhood in the year after September 11 that at one meeting that summer, when I introduced myself, a member of the group (who was also active on the Community Board and in other organizations) asked the group's leader, "So I just want to know. Is it now the group's policy to let every academic in to study us?" He felt fellow residents were allowing themselves to be put on display. At the time I had no response, already feeling like a rubbernecker on the highway who slows to view an accident scene. I had come to Battery Park City in the midst of an ongoing tragedy, to watch. As more time passed and my investment in the community grew, I no longer felt like an interloper. But in the beginning, residents felt they were under a spotlight, and I felt intrusive even being there.

I had begun with the intention of writing a critique of it as a privatized space and faux-urban citadel, expanding on the postmodern urbanists' critique with observations from an in-depth study of the place. But as I got to know the residents more fully and personally, the aggressiveness of the critique became both less supported by what I learned and more difficult to level at a social structure that was composed of actual people I knew and liked and whose opinion mattered to me. It was as I became enmeshed in my field research that I realized that ethnography had almost always been used by urban sociologists to study disadvantaged people they wished to valorize, not privileged people they sought to criticize, and for good reason: such an intimate method is ill-suited to invective.

My critique of Battery Park City (and this book remains a critique of it, and of the inequality rooted in the spatial organization of global cities that Battery Park City represents) was further complicated by the conflicted nature of Battery Park City's recent history: while it is a citadel of global capital's wealth and power, it is also a neighborhood that has suffered one of the greatest acute tragedies in recent U.S. experience. Residents suffered on September 11 and for a long time afterwards, and a community that had otherwise been exceptionally fortunate and privileged had to struggle—though *struggle* is a word rarely used for communities with incomes nearly three times their city's median—to prevent its own destruction, to reestablish community ties, and to have its voice and preferences heard above those of even more powerful actors. Yet testing outsiders' sympathy, even during that earnest struggle the community sought to keep excluding other people from their gilded enclave. Thus the criticisms in this book exist in tension with sympathy, both justified and not, toward residents. Rather than delete either that empathy or that criticism, I hope that the result is the most honest portrayal possible: one that presents the community and the problems it has faced, while insisting that the neighborhood was initially structured by plans that sought to continue economically and racially polarizing the city, and that residents have acted to reinforce rather than challenge that separation.

The fully rounded portrait that ethnography forces upon the researcher means that this book is still not as singularly critical as many other books written about the global elite and the polarization of New York City. I hope that readers, rather than faulting me for seeming to pull my punches, will appreciate that this view of an elite community is less polemically useful but still valuable—even for advocates of social change—to the extent that it provides a closer, more complicated portrait of a community at the opposite end of the economic spectrum from those that progressive activists work in solidarity to strengthen.

This project presents both the challenges facing a strong, supportive community and a critique of the elite, inequitable, undemocratic approach to building today's city that the community represents. Employing the model of the reciprocal relationship replaces a static view of space with a dynamic understanding that highlights both the role of elites in shaping space and the ability of that space to influence the actors who will occupy it. The reciprocation between actors and space over time underscores that, even for a relatively new community, an ethnography benefits immensely from knowledge of that community's history. To that end, chapter 1 sets the stage with an examination of Battery Park City's origins and the first cycle in the relationship that would shape the neighborhood, its residents, and the city for decades to come.

1

Creating Battery Park City

Building a Landmark on Landfill

BETWEEN 1960 AND the present, New York changed from a working-class, industrial city of specialized manufacturers to a global city more singularly cast than ever as a command-and-control center of global capitalism. In the same period, American cities went through equally dramatic social transformations. Places like New York experienced convulsions of population loss as whites fled to the suburbs and African Americans migrated from the South to become majorities or pluralities in many northern cities. New York went through a wave of disinvestment by businesses, banks, and landowners, then experienced a flood of new money as those same players saw the potential to profit from the gentrification of New York as an elite global city. During this period elites vacillated between positive and negative attitudes toward cities like New York, seeing the metropolis first as an industrial economic engine, then as a landscape whose design was physically obsolete, then as a cauldron of racial change and social protest, and still later as a promising site for profitable recolonization.

The fifty-year transformation of industrial Hudson River piers into the luxury enclave of Battery Park City encapsulates this spatial, economic, and political reorganization of New York City so well that it provides an ideal perspective on the successive visions for restructuring New York that elites have implemented. More specifically, earlier work has shown that the physical designs that elite real estate developers and city builders choose for large construction projects like Battery Park City are uniquely positioned to tell the story of this complex urban transformation.[1] In fact, projects such as the master plans for Battery Park City are leading indicators of the shape of cities to come, offering some of the earliest evidence of elites' plans, not just for a single project, but for the city at large.

New York's public spaces went through four distinct stages in the last half of the twentieth century. This chapter explores them by examining the succession of master plans for Battery Park City, explaining the history that gave Battery Park City its physical shape and social configuration and demonstrating why contests over the shape of Battery Park City reveal changes in the whole city, not just one community.

Public space consists of parks, plazas, malls, sidewalks, and street corners, all the generally accessible parts of the city where one is likely to meet and interact with strangers.[2] Each of the four stages of New York City's history over the latter twentieth century has been characterized by a particular type of public space. Through the 1950s, New York was a city of *working-class spaces* like the piers. By the early to mid-1960s, elites' reaction to civil rights struggles, population decline, increases in crime, and their own insecurities was to create *privatized spaces*. These included designs for development projects and public spaces that would barricade them from the larger public with fences, walls, elevated platforms, and locked gates. In the third stage, elites sought to selectively recolonize the city through *filtered spaces* such as shopping malls and enclosed atria that would be attractive to an upper middle class that developers sought to draw back into the city, but would keep out working-class and poor people and people of color. Finally, consistent with the ultimate shape of elites' gentrifying program of urban restructuring, under its last master plan Battery Park City developed the strategies of *suburban spaces*, creating lush environments for upper-income residents and employees and using distance to segregate disadvantaged people from those resources.

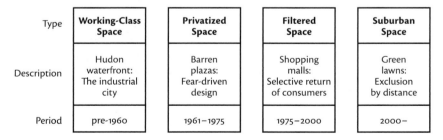

Type	Working-Class Space	Privatized Space	Filtered Space	Suburban Space
Description	Hudon waterfront: The industrial city	Barren plazas: Fear-driven design	Shopping malls: Selective return of consumers	Green lawns: Exclusion by distance
Period	pre-1960	1961–1975	1975–2000	2000–

Typology of urban space. Typical spaces constructed by structural speculators in each of the periods shown above shared key traits.

A History of Exclusion

In the context of discussing Battery Park City, the word *exclusion* has two overlapping meanings. First, it refers to the decision to price housing in the neighborhood at luxury rates so that lower- and middle- income people generally cannot live there. Second, according to some critiques, it refers to exclusion through the neighborhood's design: planners have surrounded the neighborhood with physical barriers—a difficult-to-cross highway, guard booths, hard-to-find entrances—that have the effect, on the ground, of keeping away people who are not wealthy, either because those obstacles make it difficult to enter the neighborhood or because stylistic choices make people of modest or normal means feel they are not supposed to be there. Critics argue that these designs also cocoon residents in attractive but underused outdoor spaces that lack a broad public character, leaving residents without opportunities for democratic public interaction.

Architects distinguish between a "program," which describes the functions that a client wants a building to accommodate, and a design, which is the form the building takes while serving those functions. Thus the "citadel critique"—that Battery Park City was intended to function as a refuge for rich people—describes what I call *programmatic exclusion*, while the claim that the neighborhood has been designed to use physical barriers, like the daunting West Side Highway, guard booths, or hidden entrances, to keep out the rest of us describes what I call *design exclusion*.

The social privilege that elites sought to reinforce in Battery Park City produced the neighborhood's programmatic exclusion, while a shifting social context and a long history of failed master plans eventually relocated the focus of design-based exclusion efforts beyond the boundaries of the neighborhood to New York's low-income communities of color. The ultimate design, as modified through generations of master plans, proves critical to understanding the community that came to live there.

1962: The Plan for a Working Port

Before becoming the contested site for a major development project, the area that would be Battery Park City had been a working port almost since the city's founding by the Dutch.[3] In the mid–twentieth century, the piers needed to be repaired, and this prompted the original plan for the Battery Park City area.

The city's Department of Marine and Aviation, which oversaw the piers, proposed maintaining the area's role as a working port. Warehouse

Conditions before the construction of Battery Park City shaped the neighborhood, including pierhead and bulkhead lines, Hudson River train tunnels, and the condition of existing piers. (Wallace K. Harrison, "'Battery Park City': New Living Space for New York, a Proposal for Creating a Site for Residential and Business Facilities in Lower Manhattan, 1966," Wallace K. Harrison Archives, Avery Library, Columbia University)

and trucking facilities would operate at ground level, with high-rises surrounded by elevated plazas built above. The project would include 4,500 apartments, a hotel, and office space.[4] Though this plan made room for white-collar workers, they shared priority with working-class longshoremen, sailors, and Teamsters.

The 1962 plan was the last acknowledgment in planning documents for Lower Manhattan of the city's working-class base. In that era, New York had more manufacturing jobs than Philadelphia, Detroit, Los Angeles, and Boston combined.[5] In the decades after the end of the Second World War, 25 to 30 percent of working New Yorkers were unionized, with over a million workers paying union dues. The city's largest union, the International Ladies' Garment Workers' Union, had nearly two hundred thousand members into the 1960s. Nonetheless, new economic and real estate development in New York was not directed toward the city's growth as an industrial center. After the plan by the Department of Marine and Aviation, business executives, real estate developers, and government agencies drawing up plans for Battery Park City never again considered warehouses, ferries, cargo ships, or commercial produce markets in their plans. For this reason, the sociologist Sharon Zukin calls this era the "threshold period," when industry began its move to suburban locations.[6] Within a year, decision makers' plans for the area that would become Battery Park City would definitively shift their focus to the white-collar financial sector of the city.

The 1962 plan made little headway, but it introduced several important ideas that would ultimately shape Battery Park City. It was the first to recommend a mix of office and residential uses for the space. It signaled to others the city's recognition that significant space near Lower Manhattan needed to be refurbished and suggested to developers that it might be up for grabs. For some time elites' plans had recommended uprooting New York's working-class, specialized industrial base and replacing it with a city serving the immediate needs of financial firms.[7] Downtown decision makers soon proposed transforming the waterfront from blue-collar to white-collar uses. Many writers have described Battery Park City, as it was ultimately built, as the product of global capitalism. But rarely do we see how such a global force shapes a specific, local area. The first white-collar plan for Battery Park City presents the individuals whose actions interpreted *abstract and global* social changes—in this case, the rise of global capitalism —in the shaping of *concrete and local* spaces like Battery Park City.

1963: Chase Manhattan Bank and David Rockefeller's Plan

Manhattan has two central business districts: the Financial District, on the site of the original European settlement on Manhattan at the southern tip of the island, and Midtown, three miles north in a part of the city organized by the well-known grid of numbered streets and avenues.[8] For decades, Downtown's elites had worried that corporate headquarters would continue to abandon Downtown for Midtown because of the competitive disadvantage of the Financial District's restricted space and smaller lots.

The Battery Park City project was proposed in the early 1960s by David Rockefeller, then vice-chair of Chase Manhattan Bank, to retain and attract corporations to Lower Manhattan. Rockefeller had recently led Chase in their decision to build a new headquarters in the smaller, more crowded and economically faltering Downtown rather than in Midtown. First National City Bank, Chase's major competitor, had moved to Midtown just four years before the new Chase building was dedicated, and Chase's considerable investment motivated the bank to promote a general revitalization of Downtown.[9] At the suggestion of the city builder Robert Moses, Rockefeller founded the Downtown–Lower Manhattan Association (DLMA) in 1958 to "speak on behalf of the downtown financial community and offer a cohesive plan for the physical redevelopment of Wall Street."[10]

Rockefeller was particularly well suited to lead such an effort. He was the youngest of the six children of John D. Rockefeller Jr., who, as heirs to the fortune of their robber baron grandfather, had had to develop public identities in relation to their private wealth. David's brother Nelson had gone into politics; David instead built a more stable network of power in business, becoming chief executive at Chase Manhattan Bank and developing extensive international contacts among businessmen, dictators, and politicians.[11] (The head of Goldman Sachs once complained, "David's always got an Emperor or a Shah or some other damn person over here, and is always giving him lunches. If I went to all the lunches he gives for people like that, I'd never get any work done.")[12] As with the DLMA, Rockefeller typically promoted his goals by leading a larger coalition of like-minded businessmen.[13] The DLMA was not the only organization to propose renovating Downtown. But given his enthusiasm for such projects and the influence that accompanied the power and wealth of Chase and his family's fortune, it is not surprising his organization's plans set Battery Park City in motion.

Observers of global cities argue that global capitalism produces citadels as protected enclaves for capitalism's elite, but it has not been clear how two distinct systems such as global capitalism and, for instance, New York City urban planning ever actually connected to each other. David Rockefeller was a universal joint that connected the former to the latter. He was proudly globalist in outlook, writing, "Some even believe we are part of a secret cabal . . . characterizing my family and me as 'internationalists' and of conspiring with others around the world to build a more integrated global political and economic structure. . . . If that's the charge, I stand guilty, and I am proud of it." But his interest in the planet was not altruistic; his goal was to increase U.S. control of the international economy and of resources in other countries. His interest in Latin America led him to criticize a Kennedy-era proposal, for instance, "for not insisting strongly enough that the Latin nations promote the expansion of private US capital."[14] He was an early backer of the U.S. invasion of Vietnam, and he provided loans to help the U.S.-allied apartheid government of South Africa after other nations withdrew investments to protest the 1960 Sharpeville massacre. While it is important to avoid the trap of attributing too much influence to a single, publicly visible figure, Rockefeller fit the mold of a global capitalist and had the influence to lead elected and business officials to seriously consider plans to build Battery Park City. He was one of many people involved who illustrated how, exactly, globalization constructed local citadels.

The DLMA immediately issued a report in 1958 calling for improved infrastructure, changes in land use, major building projects, and demolition of the Hudson and East River piers to make room for the needs of the DLMA's financial industry members.[15] The DLMA's 1963 report was the first to propose Battery Park City in recognizable form, as a mile-long landfill project for luxury high-rises. The plan for Battery Park City did not arise in isolation, however. It was one element of the overall plan of the DLMA and Downtown interests to enlarge and modernize Lower Manhattan by providing much more land for office and apartment construction. Thus Battery Park City was one of eleven major improvement projects recommended in the 1963 report, and the World Trade Center was another. The DLMA had actually already proposed, in 1960, both a World Center of Trade (or World Trade Mart) and a cluster of high-rise housing near the old Battery at the southern tip of the island, but had sketched them out on the East River, on the other side of Manhattan.[16] The Port Authority (which answered to New York and New Jersey's governors) had agreed to the building of the World Trade Center, but at New Jersey's insistence it stipulated that the site be moved from the East River to the Hudson River, above the terminus of the New Jersey–New York commuter train line that the Port Authority had just taken over. When the Trade Center moved, Battery Park City moved with it to the piers New York had already been planning to redevelop.[17]

The ambitiousness of the DLMA's plans—reconfiguring all of Lower Manhattan in an effort to "modernize" it—was consistent with the scale of city building in that era. New York's notorious power broker Robert Moses, who was building highways, parks, and high-rise housing and was bulldozing thousands of homes to do it, approved of the DLMA's vision. He sought support for his own Downtown project, an expressway across

The 1963 plan for Battery Park City by David Rockefeller's Downtown–Lower Manhattan Association was the first to link Battery Park City to New York's role as a global financial capital. Several elements shown here were actually built: the World Trade Center, Battery Park City, the Civic Center government buildings, and the East River heliport. (Downtown–Lower Manhattan Association, *Major Improvements: Land Use, Transportation, Traffic* [New York: Downtown–Lower Manhattan Association, 1963], 6)

Lower Manhattan that would have cut through SoHo, Chinatown, and Little Italy, with these words: "Not to be overlooked in examining the benefits of the Expressway is the tremendous stimulus it will provide to the program of the Downtown–Lower Manhattan Association. . . . This group of distinguished downtown leaders, headed by David Rockefeller, is assiduously tacking a giant task—the rehabilitation of Lower Manhattan. These citizens have undertaken a project of stunning scope. They deserve nothing less than the complete cooperation of our elected and appointed city officials."[18] Moses recognized how his work and that of the DLMA were reshaping the city. Visions of the "rehabilitation of Lower Manhattan" were the first inklings of the new global city.

To reshape the city in this way, the DLMA rejected the Department of Marine and Aviation's earlier mixed-use plan for a working port and instead proposed using the whole site for luxury apartments. In the DLMA's illustration, rows of high-rises march down the length of the site. The group hoped that high-end housing would make Downtown more attractive and convenient to financial firm executives. From its first inception, Battery Park City was intended to be part of an engine to revitalize the Financial District, but only by providing luxury housing for executives who worked in the area—not by providing housing for the more numerous lower-paid pink-, blue-, and white-collar workers in the those same companies, even though the accessibility of a potential workforce also influences companies' location decisions.

Moses endorsed the DLMA's plans, but because his projects consisted largely of government-funded urban redevelopment and highway build-ing, they were less singularly devoted to the financial industry elite. In this respect the DLMA diverged from Moses's best work. Moses constructed many projects on the scale of Battery Park City, tearing down blocks of apartments to build projects like Lincoln Center on a tabula rasa much as the DLMA envisioned landfill on the Hudson. He had no more regard for the needs of the low-income, often black and Hispanic people he displaced than the DLMA had for those same groups in the work they did. But be-cause many of Moses's projects constructed parks and basic infrastructure, his projects built things—bridges, highways, beaches, housing—that were widely used, and therefore had a populist element. I offer this observation not in the spirit of some recent efforts to rehabilitate Moses's reputation, but to point out that from its inception Battery Park City represented a shift in strategy. Moses's top-down modernist plans to rebuild the whole city gave way to plans to build only for elites. Battery Park City retained Moses's attraction to megaprojects but dispensed with even the authoritar-ian populism that his projects often carried.

The genesis of plans for Battery Park City is significant because although the citadel hypothesis requires that the ideology of a global economic sys-tem physically manifest itself at the micro level of individual city blocks, parks, and buildings, *how* that ideology can be transmitted to local urban forms is not immediately evident. The role of the DLMA demonstrates how such transmission occurs: self-conscious elites of the burgeoning global financial industry, working to protect their real estate investments and strengthen their company's strategic position, collaborated to propose that city, state, and local governments invest in reshaping the Downtown for the industry's benefit. Rockefeller, while only one member, was a founder, and it was precisely his elevated position in the hierarchy of global capitalism that helped the plan move forward. As CEO of Chase Manhattan, he spoke for a company that according to historian Robert Caro was once "very probably the single most powerful financial institution on the face of the earth."[19] His experience and connections in government and his brother's position as governor of New York further facilitated gaining an audience for the plan. And his identity as a self-avowed "globalist" who sought to better integrate global political and economic structures provided him the vision to position New York, and his bank, as a central node for global corporations.[20] In this way, this early proposal for Battery Park City, while it offered scant details on the design of the neighborhood, provides a useful understanding of the real ways large-scale economic systems link to local city-building projects.

Just as the Marine and Aviation plan contributed durable ideas to the Battery Park City vision, so the DLMA plan shaped the decades of plans that followed. Most importantly, it eliminated industrial functions from the port for the first time and oriented the project completely toward the financial industry functions of Lower Manhattan and the needs of high-income, white-collar workers. The DLMA's vision of an elite Battery Park City would never be displaced.

1966: Privatized Space

By the mid-1960s, New York's racial changes were evident, as African Americans and Latinos (largely Puerto Rican migrants and later immigrants from the Dominican Republic) moved to the city and as whites (including many second- and third-generation Irish, Italian, and Jewish Americans) accelerated their move out to the suburbs. Simultaneously, whites deduced from the civil rights movement that the old rules of racial deference, de facto disenfranchisement, and unassailable inequality could no longer be taken for granted.[21] With the dismantling of the old racial order, whites anticipated disorder, and racially coded tales of crimes reinforced that sense and the belief that the city was becoming a dangerous place. As Derek Edgell wrote, "Some of the incidents involving attacks on whites and Jews served to create in the minds of 'worried' elements in the city an association between blackness and criminality which, although simplistic in the extreme, was real enough for them. Anxieties about young ghetto males being out of control began to take on exaggerated form after 1963."[22] The privatized design of public spaces and development projects from this period reflects those anxieties.

Two books directed at privileged New Yorkers stoked the sense that racial change had unleashed chaos in their city. Eric M. Javits's *SOS New York: A City in Distress* was not subtle. Making clear his intended audience,

Javits's first example of random, uncontrolled crime in the city was of "a businessman's wife . . . stabbed by an assailant while walking her dog in broad daylight in Central Park near the Metropolitan Museum of Art."[23] Describing *her* was less important than identifying her as a *businessman's* wife. There was plenty of crime in the city, but it was a sign of disorder that it should befall this type of person, in this kind of place, at this time. Similarly, Richard J. Whalen's *A City Destroying Itself: An Angry View of New York* began as an essay in *Fortune* in 1964 and was addressed to typical readers of that magazine. To write the book, Whalen easily summoned his "outrage" "from the bountiful store that every New Yorker possesses, but habitually suppresses."[24] While air pollution and traffic were sources of Whalen's irritation as well, the "mounting disorder" he wanted to warn readers about had strong ties to racial change and a concomitant sense of growing insecurity. "The city itself sways on the edge of madness. It almost plunged over during the summer of 1964, when Negro mobs surged through Harlem and the Bedford Stuyvesant section of Brooklyn, rioting and looting." Whalen claimed there were "mere children" in the mob, "already filled with contempt for 'Whitey's' law."[25] (He made no mention of James Powell, another "mere child," whose murder by a white policeman had sparked the riots.)[26]

Such fear manifested itself in public space. In 1963, a *New York Times* article announced, "Fear Said to Keep City Parks Empty: City Group Says Its Survey Shows People Are Afraid Even during Daylight." The article was based on a survey of park users by the Park Association. While they had planned to determine users' opinions of "shrubbery, benches, fountains, pavement and the like," the association found that people's overriding preoccupation was fear.[27] While the Parks Department tried later that year to rebuff charges that the parks had become unsafe, the tide had clearly turned: even the police deputy inspector who issued a report claiming that, contrary to the Park Association's findings, crime had declined, seemed unable to maintain an image of public safety. He admitted, somewhat contradictorily, that arrests had risen, and he conceded that he wouldn't want his wife and daughter to enter a park at night without his protection.[28] Meanwhile, a *Times* reporter made exaggerated "suggestions that the park might be unsafe for *unarmed* or unwary pedestrians."[29] Fear of public spaces spread from the people sitting on the benches to the cops patrolling the parks. Such widespread fear—even during daylight, and even though parks, then as now, had lower crime rates than the rest of the city—affected the daily patterns of regular New Yorkers and the design plans of decision makers' major development projects.

From the 1966 Harrison master plan, a beautiful illustration of spaces that probably, if contemporaneous developments are any indication, would never have had the teeming life pictured on these immense, elevated plazas. Notice the continuation of the older street grid from beyond the elevated highway. (Wallace K. Harrison, "'Battery Park City': New Living Space for New York, a Proposal for Creating a Site for Residential and Business Facilities in Lower Manhattan, 1966," Harrison Archives, Avery Library, Columbia University)

It was in this context of growing anxiety about public space that the first state designs for Battery Park City were drawn up. Having initiated the project, David Rockefeller and the DLMA played a much less active role in promoting it from then on. By that time, however, Battery Park City already had gained proponents among state and city politicians, including Governor Nelson Rockefeller and Mayor John Lindsay.

The Battery Park City master plan drawn up in 1966 by Wallace K. Harrison for Governor Nelson Rockefeller was ambitious and impressive. The beautifully illustrated, detailed plans show apartment blocks separated by brilliant white plazas and neat groups of trees. People wander around marinas, along waterfront walkways, and in glass-fronted stores and other facilities, watching sailboats on the river. In the model, the World Trade Center stands guard behind the project, while in Battery Park City itself, the modernist apartment blocks are bookended by sets of skyscrapers at the north and south ends of the project. But close examination of the plan challenges several assumptions about the inclusiveness of plans from the early years, in both their design and their intended income mix.

Conventional accounts of the development of Battery Park City describe this plan as part of the fabric of the more inclusive Great Society programs, in which the government would build new housing for the poor, working, and middle classes. The current account by the Battery Park City Authority (BPCA) is typical when it describes the 1966 plan as exemplifying the "Integrated Society": "[Governor] Rockefeller needed to house people coming out of slums, and Harrison's plan was a reaction against the appalling conditions of tenement housing."[30] Yet once the numbers are untangled, only 10.6 percent of the plan's 13,982 units were for low-income residents, compared to 46.8 percent for "upper-income" residents.[31] Far from envisioning an integrated society to replace slum tenements, the plan provided for a segregated community consistent with both the DLMA plan and what was eventually built: upper-income housing for elites near the Financial District, with a smattering of subsidized lower-income units.

Governor Rockefeller's plan exhibited not only this programmatic exclusion but the kind of dramatic design exclusion characteristic of the privatized period as well. It envisioned a neighborhood far more physically exclusive than today's Battery Park City. While there are several playgrounds in Battery Park City today, in Harrison's illustrations much of what would otherwise be public space is *inside* the high-rise towers. Inserted into various floors among the apartments are play spaces for young and older children, community areas, even "conversation areas for adults."[32]

The buildings themselves are elevated on the kind of immense, raised concrete plazas that were by that time already recognized for their ability to make a space uninviting and rarely used by people. One illustration shows access to these elevated spaces via a pedestrian ramp completely cut off from anything else, rising above highways on either side, with the nearest building blocks away. The plan would have made Battery Park City very difficult for potential nonresident users to access.

In these ways, the 1966 design of Battery Park City represents the period from roughly 1961 to 1975 in the city's history, when large projects were built according to models of privatized space. Designs from the privatized-space period typically retreat from the public life of the city, much as in white flight whites retreated from the city to distant suburbs. Buildings from this period took a step back from the street. New high-rises were surrounded by vast, empty, often barricaded no-man's-lands. Or public spaces were elevated and gated, imagined to be for the exclusive use of the residents of such luxury spaces. (In practice such isolated, elevated plazas were rarely used by anyone.) The indoor play spaces and elevated ramps, decks, and plazas of the 1966 plan were just such privatized spaces.

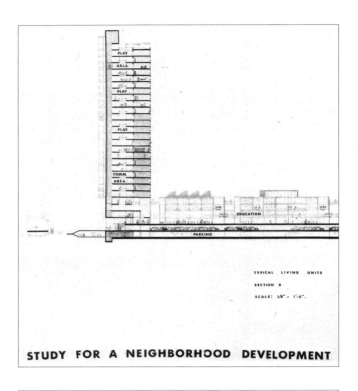

TYPICAL   LIVING   UNITS

SECTION B

SCALE: 1/8" = 1'-0".

STUDY FOR A NEIGHBORHOOD DEVELOPMENT

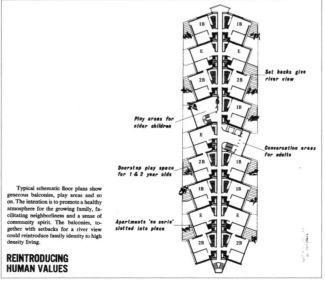

Set backs give
river view

Play areas for
older children

Conversation areas
for adults

Doorstep play space
for 1 & 2 year olds

Apartments 'en serie'
slotted into place

Typical schematic floor plans show
generous balconies, play areas and so
on. The intention is to promote a healthy
atmosphere for the growing family, facilitating neighborliness and a sense of
community spirit. The balconies, together with setbacks for a river view
could reintroduce family identity to high
density living.

REINTRODUCING
HUMAN VALUES

Above: Access to Battery Park City in the 1966 Harrison plan would have been exclusive, if anyone at all had tried to enter via this protracted and isolated ramp overlooking highways. (Wallace K. Harrison, "'Battery Park City': New Living Space for New York, a Proposal for Creating a Site for Residential and Business Facilities in Lower Manhattan, 1966," Harrison Archives, Avery Library, Columbia University)

Opposite page: Harrison's 1966 plan, like others preceding the 1979 master plan that was actually implemented, was far more private. Here, traditionally public elements, like play and community space, have been moved inside tall apartment buildings, totally insulating them from the public. (Wallace K. Harrison, "'Battery Park City': New Living Space for New York, a Proposal for Creating a Site for Residential and Business Facilities in Lower Manhattan, 1966," Harrison Archives, Avery Library, Columbia University)

Privatized space became the norm for large development projects throughout the 1960s. Most office plazas in Midtown were privatized.[33] Architects I interviewed who were active at the time reported that developers were clear in their instructions: they ordered that fewer and fewer amenities be provided in the plazas to make them uninviting to passersby. Though the rest of the city was constrained by the built environment that was already in place—a tight grid of city streets and sidewalks, existing buildings from earlier periods, and a mix of different types of workplaces and residences that gave New York street life its diverse feeling, a tabula rasa like the landfill under Battery Park City was a stage on which leaders within the "growth machine" coalition could present their most uninhibited ideal of the city.[34]

The privatized tendencies represented elsewhere in the city took on even more dramatic form in these plans.

In these early years, the city and state competed for control of the development project. Thus the same year that the governor unveiled the 1966 Harrison plan, the city produced a plan of its own, a mass of brutalist concrete buildings.[35] Interestingly, the 1966 state plan's difference from the city plan of that year highlighted the same shortcoming that would become evident in the state's plans for rebuilding West Street after September 11, 2001: building for auto traffic and ignoring the pedestrian needs of the nation's densest city. Whereas in 1966 the city plan considered pedestrian corridors across lower Manhattan, the state plan treated pedestrians only as something that could slow down car traffic. Only one hundred copies of the city's plan were produced in a medieval process of hand-typing and hand-binding the elaborate document, so the plan's direct influence was limited. Ultimately, the state would take over Battery Park City, and the city would never again develop separate plans for the site.

1969: Filtered Space

Thousands cheered as politicians praised the new South Street Seaport on its opening day in the summer of 1983.[36] A mall with a collection of themed restaurants, novelty stores, and small vendors selling unusual snacks and offbeat products, it was designed to make shopping a leisure activity.

Built on a pier over the East River waterfront, South Street Seaport was on the side of the Downtown waterfront where David Rockefeller's organization had initially proposed Battery Park City. The project was the latest in developer James Rouse's famous series of urban "festival marketplaces." Starting with the renovated historic Faneuil Hall in Boston and Harborplace in Baltimore, these malls were intended to make "downtown popular again for the middle-class," as Rouse's most appreciative biographer, among many others, described the plan.[37]

Rouse was already well known for game-changing projects, and these festival marketplaces were among his most high-profile undertakings.[38] The marketplaces defined early 1980s urbanism, which held that the goal of urban regimes was to lure the middle class back to cities—if only for an afternoon—after their two-decade-long flight to the suburbs. The malls landed Rouse on the cover of *Time* magazine in August 1981 with the iconographic headline "Cities Are Fun!" Mayors around the country scrambled to get a Rouse-style festival marketplace built in a rundown sec-

The South Street Seaport mall

tion of their waterfront. Critics pointed out that the jobs created were low-wage, part-time service industry positions that were a paltry replacement for the family wages being lost to deindustrialization and deunionization. Further, the projects brought significant government subsidies to a recreational center for the upper middle class but provided no direct aid to the city's poor who needed it the most. The goal, boosters countered, was to draw the (putatively white) middle class's disposable income back to the city, followed by their businesses, homes, and tax base. A rising tide would lift all ships.

This filtered strategy was first implemented in the mid-1970s. Rouse's malls epitomized the effort to welcome back the middle class. His first festival marketplace, Faneuil Hall, opened in 1976. But unrealized plans to develop enclosed private retail space and create a secure urban experience for the upper middle class predate his work. Battery Park City was one example.

The mall master plan was produced in 1969 by the BPCA and was included as part of the city's 1969 citywide plan.[39] Passageways forty stories in the air connected the hexagonal office buildings that were clustered at the southern end of the landfill. Apartments and the office towers were to be

The 1969 shopping mall plan: futuristic hexagonal buildings (*top*) were
to be connected by a shopping mall promenade and monorail (*bottom*).
(New York City Planning Commission, *Plan for New York City: A Proposal*
[New York: Department of City Planning, 1969])

connected to a long shopping mall of cascading levels. The mall would be
glass covered, sunlit, and lush with greenery. A monorail running through it
would connect the apartments and office buildings. This was an ambitious
vision of Battery Park City as an all-purpose attraction for professionals in
the Downtown financial sector. Like Rouse's malls, which would soon become icons of the age, this project abandoned earlier privatization efforts
to close off public space and instead sought to reverse the flow of affluent
customers to the suburbs by rebuilding and carefully guarding an engag-
ing retail space under private management. The "filtered space" period of
American urban history was under way.

Unlike privatized spaces, which are unwelcoming to everyone, filtered spaces welcome some (typically more affluent) users, while discouraging use by others and employing strategies to control the behavior of those who do use the space. In addition to the Rouse malls and Battery Park City's early mall plan, after zoning changes in 1975, growing numbers of office plazas in Manhattan were filtered, mall-like spaces. While such spaces could be undeniably fun for some users, and could even feel "public" and unrestricted to them, others entered with more trepidation. The goal of such spaces was to create the conditions for a selective return to Downtown by whiter, more affluent customers, workers, and residents. Filtered spaces were an early step in gentrification and the large-scale reappropriation of center-city New York for upper-income people, homes, and businesses. Filtered spaces were examples of more thoughtful, vibrant urbanism but were nearly as exclusive as their predecessors.

The 1969 plan is also significant for its intended income mix. Contrary to the "rising tide" approach that most such mall projects took, which was to neglect low-income residents, the 1969 plan was developed in the one brief period when plans dictated Battery Park City would be more diverse. This proved a short-lived anomaly. Mayor John Lindsay had maintained until April 1969 that *all* of Battery Park City should be for high-income residents. But by 1969, liberal Democrats in the city were at the height of their influence and demanded more low-income housing in the Battery Park City project. Lindsay, running for reelection, needed the support of liberal politicians, so he endorsed a new mix in which low-, middle-, and upper-income residents would each have a third of the units. (Some groups at the time dismissed this as "the kind of tokenism which is an affront to the needy." Since there are numerically fewer upper-income people and they have more housing options, even the numerically even split was not an equitable one.)[40] But no action was taken to realize the plan, and in less than three years the BPCA had successfully lobbied for a rewritten lease that brought back the kind of luxury-laden income mix that had been the rule all along.[41]

It wasn't that a high-income neighborhood was the only feasible option; rather, it was the only option that decision makers ever seriously pursued. In the procession of plans for Battery Park City in the past forty years, poor residents have almost always been expected to make up no more than 7 to 14 percent of the population.

There is more evidence that promoters of Battery Park City never intended it to be anything but an upper-income community. For his 1976 dissertation, Maynard T. Robison conducted extensive interviews with

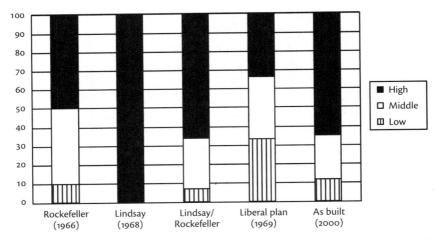

Varying proposed income composition of Battery Park City in 1960s plans and actual income composition of the neighborhood in 2000. Though it was forty years in the making, and though nothing was built for two decades, the income mix looks most like that proposed by Rockefeller and Lindsay in 1966. Because of the vagueness of "low," "middle," and "high" categories, figures for today are approximate: "low" describes those households making under $25,000; "middle," those from $25,000 to $75,000; and "upper," those above $75,000.

leading players in the earlier development of plans for Battery Park City. Robison found that "for Lindsay, Rockefeller, and others deeply involved in planning Battery Park City, high-income housing was the only proper use of the shore area."[42] Further, Robison concluded that Battery Park City's elite demographic was not caused by construction delays that deferred the project until affordable housing funds had dried up. Rather, the reverse was true: elites' resistance to building affordable housing led them to forestall any action on the project during periods when populist politicians could require more affordable housing. According to Robison, "The decade of delays in Battery Park City's construction resulted, primarily, from a single cause: the desire of virtually all those deeply involved in the project to build it as a high-income enclave."[43]

Evidence in the master plan supports this conclusion. The narrative of tight financial times would suggest those lower-income units were replaced by luxury units, but the current plans suggest otherwise. The most inclusive master plan was for 14,100 units, 9,446 of which would have been upper-middle-class and upper-class units, the remaining third of which would have been low income. In the end, the Authority planned to have 9,664 units built, almost all for the luxury market. Thus low-income units were not replaced but cut out of the plan. The commercial complex was always expected to subsidize the residential component, but while the BPCA does end each year with a surplus, it has done relatively little to build affordable housing on site.

Robison also argued that because "issues of class and issues of race are almost always intermixed," and because "economic segregation is far more acceptable than proposing racial segregation," plans for an economically elite neighborhood were essentially plans for a white neighborhood in an increasingly nonwhite city facing considerable white flight. According to Robison, there was

> a good deal of evidence that what they were really after was racial segregation, an all-white project. Virtually everyone involved conceived of the low- and, to a lesser extent, moderate- and middle-income housing they didn't want in lower Manhattan as housing exclusively for Blacks and Spanish-speaking people, "luxury" housing as housing for whites (although only a very small proportion of the city's non-Spanish-speaking white population could afford the luxury rents). The most striking evidence of their attitude was offered by the statement of one of the major actors. "No one would mind if Ralph Bunche lived there." Unfortunately, Bunche, the former deputy secretary general of the United Nations and possibly the one Black perfectly acceptable to whites, had died some time before this statement was made.[44]

Battery Park City's racial composition today is much what Robison, writing before anything was built in Battery Park City, expected from interviews with decision makers.

Little of the 1969 plan survived. The income mix was scrapped. The luxurious and futuristic mall design would not have fit with the economics of affordable housing, or even luxury housing. Orienting everything along a mall and monorail meant the plan required substantial infrastructure investments. Developers were not interested, and the plan never got off the drawing board.

By this time, across the street another mall was planned that would come to fruition. The World Trade Center represented a transitional form between privatized plazas and filtered shopping malls. It was designed and built in a period that spanned the privatized and filtered eras: plans for the

World Trade Center had shifted to the West Side, above the train lines to New Jersey, in 1961. Demolition of the buildings on the site began in 1966, and the towers were dedicated in 1973.[45] At street level, it looked privatized on most days, a huge plaza that offered few amenities to the public.[46] The plans included an immense underground mall, but hardly one that could attract visitors the way South Street Seaport would in the 1980s. Instead, it was designed to provide amenities for the small city of workers in the offices above, much like the mall at the Citicorp building uptown. Despite its origins as a privatized space, the plaza and lobbies of the Trade Center were later adapted into part of the built environment that filtered access to Battery Park City. The most common means of reaching Battery Park City was to enter the lobby of one of the Trade Center buildings, which also served as the entrance to the pedestrian bridge leading to the Winter Garden Mall. That passage was effortless for the white-collar workforce of Lower Manhattan, thirteen thousand of whom traveled across the bridge each day to commute from New Jersey via ferries that docked in Battery Park City. But people who did not feel welcome in office building lobbies found the entrance more intimidating, a key feature of filtered space.

Batter Park City's planners would eventually adjust their designs in response to their new neighbor, relocating Battery Park City's office buildings next to the Trade Center. Residents and workers in Battery Park City would make more extensive use of the Trade Center once the Trade Center station became their closest subway stop, the mall became where they found a pharmacy, a bank, or got photos printed, and the bridge became the entryway to their neighborhood. But the social impact of the World Trade Center, a project whose financial prospects, identity, and success remained uncertain for decades after its construction, would remain unclear until Battery Park City was largely built.

1975: The Continuation of Filtered Spaces as Exclusive Pods

Filtering did not only take the form of designs that selected affluent customers for exclusive shopping malls. Designers also followed new guidelines that promised to filter people out of residential buildings. Battery Park City's 1975 plan applied this kind of filtering.

Such strategies were developed and promoted by planner Oscar Newman, who described the approach as "defensible space." Newman is most well known for his work on crime control in public space. His landmark book, *Defensible Space: Crime Prevention through Urban Design*, was published in 1972. *Defensible Space* and the dozen works in which Newman elaborated his recommendations established his approach as the model for security-oriented design.

For Newman, "defensible space" was composed of "real and symbolic barriers, strongly defined areas of influence, and improved opportunities for surveillance—that combine to bring an environment under the control of its residents."[47] It combined "target hardening," like stronger locks and abundant video cameras, with physical designs that put residents in positions from which to observe and control entry to their buildings and public spaces. He pointed to the importance of good visibility, entryways over which residents felt control, and buildings organized so that residents could recognize anyone else who shared their entryway. Exclusive control of a semipublic space could be demonstrated to outsiders with "changes in texture and color of paving; raised stoop . . . and overhang from balcony above." Despite its promise to make life safer for everyone, Newman's work on hardening neighborhoods against predatory criminals from elsewhere contributed to a discourse on urban safety that was already deeply racialized.

To his credit, Newman was working to address issues of basic safety that disproportionately affected people of color. Though modernists' optimism had been tempered, his work was a continuation of the belief that the design and provision of public housing could improve people's living conditions. Limited points of entry, security cameras, and highly visible play spaces were part of the reformist tradition of public housing, and he enlisted them to improve, first and foremost, the lives of poor, often minority people. Nonetheless, his recommendations regarding neighborhood safety were inextricably tangled up with race.

Consider the concept of "defensible space," which, as its name suggests, is constructed so that residents can identify and expel outsiders. Writing of New York's Co-op City, a popular but relatively isolated mixed-income development to whose safety New Yorkers from increasingly dangerous neighborhoods had fled, Newman observed a racial component to the area's safety and assumed that criminals could be identified by dark skin. "So long as all the families in Co-op City are exclusively white, middle-class, and elderly, the crime rate will stay down. The appearance of anyone else sends out a danger signal as obvious as an alarm bell. But already there are young families moving into Co-op City—black families, Puerto Rican families—seeking the same security and using the same means to achieve it. *As the population becomes mixed, the success of this strategy will diminish.*" Newman apologized, but not for long. Though he wrote immediately afterwards that Co-op City's defensive posture "will not work for very long, and it is repellent by virtue of the racism and prejudice it practices," Newman himself implicitly described criminals as nonwhite: "The lesson to be learned from Co-op City is that crime control can be achieved by creating a situation in which it is possible for the potential victim to recognize in advance the potential criminal." Evidently all criminals had looked like the new residents: young blacks and Puerto Ricans.[48] Because Newman failed to challenge the racism already permeating everyday discussions of crime and criminals, his recommendations were impaired by it as well. Defensible space sought to resegregate under a regime where explicit segregation was publicly unacceptable.

Newman's observations sought legitimacy as tough-but-honest descriptions that weren't racist because they were "real." But defensible space may have been just the opposite. Newman advocated places (presumably like Co-op City) in which an outsider "perceives such a space as controlled by its residents, leaving him an intruder *easily recognized and dealt with*."[49] But making easily identified outsiders feel unwelcome cannot be seen as a race-neutral strategy in the twentieth-century United States, and it presumably was not seen that way by the first African American and Puerto Rican families to seek security and freedom in Co-op City.

This approach allowed Newman to blame intellectual liberals' social policies and Black Liberation for crime. Like an angry New Yorker in the tradition of Richard Whalen, the author of *A City Destroying Itself*, Newman argued of neighborhood diversity that "although this heterogeneity may be *intellectually* desirable, it has crippled our ability to agree on the action required to maintain the social framework necessary to our continued survival. The very *winds of liberation* that have brought us this far may also have carried with them the seeds of our demise."[50] If the seeds of our demise were caused by intellectuals and liberation movements, then the built environment had to constitute a response to liberal permissiveness and radical challenges. In a society where images of racial minorities and crime were so intertwined, and in which Whalen also conflated the civil rights movement with street crime, perhaps it is unsurprising that liberals' proposals to desegregate were imagined as an invitation to criminals to enter white preserves.[51] How many people conflated intolerance of crime with intolerance of racial minorities? Could the designers of secure public space make such a distinction?

The planners of Battery Park City's 1975 master plan explicitly employed Newman's defensible space concepts in their design.[52] Today, the Authority titles the 1975 master plan "defensible space."[53] By then, the landfill that

Mothers push strollers in front of Gateway Plaza, the largest and oldest development in Battery Park City (March 2004).

would be the foundation of Battery Park City was complete, but nothing had been built on its sandy expanse. The 1975 plan was slightly less ambitious than earlier projects but in significant ways was more intentionally exclusive. The plan proposed "pods" that were clusters of high-rises organized around a common, private courtyard greenspace. David Gordon's generally flattering book on Battery Park City says, regarding the 1975 plan, that developer Samuel Lefrak and BPCA chair Charles Urstadt "were quite concerned about residential safety at the time, and Oscar Newman's 'defensible space' concepts were highly visible. The new pod designs were like a fortress with a single guarded entrance."[54] Each pod, home to thousands of residents, was then to be connected to others via elevated walkways. Such a design followed Newman's instructions carefully: residents were welcomed in, but the space was designed with the intention of easily seeing and identifying "outsiders."

One pod complex was actually built. Gateway Plaza, a 1,712-unit complex with three high-rise and three low-rise towers surrounding a central courtyard, was the only element of this plan to be realized and became, in 1982, the first buildings in Battery Park City.[55]

Defensible space: everyone entering one of Gateway Plaza's six towers, by foot or by car, uses a single entrance.

Residents enter Gateway Plaza through a low-roofed driveway or on narrow sidewalks to either side of the driveway. The garden in the middle is well maintained, but signs make clear that it is not for public use, and it is rarely used even by residents. This master plan, like the previous one, originally kept pedestrians and cars on separate levels, a scheme that is widely recognized today as having a deadening effect on public space and that in other studies has been found to create pedestrian zones that are exclusive and unwelcoming to less affluent potential users.[56] As the BPCA writes today of the 1975 plan, "The city at large was kept out. This was the developer's solution to the problem of middle-class suburban flight: the attempt was to bring the suburbs to the city. For a while, elevated sidewalks connected all the pods, while traffic moved around underneath."[57]

Although Gateway intentionally amplified design exclusion to create "defensible spaces," today it remains the most programmatically inclusive place in Battery Park City. It was financed with bonuses for middle-income housing, and rents there were moderated by government agreements and subsequent negotiations between the landlord and the tenants' organization. Gateway residents depended on the difference made by rent stabilization and noticed that Gateway, while not poor, was different from the rest of Battery Park City. An artist and teacher who had moved to Gateway in 2002 explained,

> There is kind of a class difference. And it's not spoken, and it's very subtle. But the people in Gateway are middle-class people who can afford middle-class apartments, [even if they're] very expensive apartments, like everywhere else in New York City. But the people with the co-ops and condos, with the carrying costs, they're outrageous. A two-bedroom [elsewhere] in Battery Park City is the same as a million-dollar place on the Upper West Side. . . . There's no behavioral distinction on the street, but I sort of feel like I'm on borrowed time. Because if Gateway Plaza leaves rent stabilization, I won't be able to live there, and most people in Gateway won't either.

The subsidies were a testament to the developer's concerns about the difficulty of renting apartments in Lower Manhattan at luxury rates in the early 1980s. Gateway's distinctive financing would make it socially important to the neighborhood in years to come, contributing a disproportionate share of the neighborhood's activists. As it eventually took shape, Gateway's economic diversity inside a defensible space design demonstrates that design exclusion does not necessarily correlate with programmatic exclusion.

The 1975 plan reflects another phase in the evolution of elites' designs for the city. Developers in the privatized era saw the city as chaotic and sought simply to barricade upper-middle-class residents from the rest of the city. The era of filtered spaces represented a more complicated conceptualization of the city. The primary preoccupation remained crime and how to keep it at bay for more privileged classes. But the concept of filtered space went along with the vision of a warmer community within the citadel, so that a hardened exterior would protect community life in the city for those few lucky enough to be inside. In this respect, the filtered period represents elites' first efforts to recolonize the city. But 1975 is a surprising time for such a strategy to be enacted. In the popular telling of New York's history, 1975 is not the year the city came back but the year it almost went under. That elites were planning a return to the city at the same time the public was expecting a final exodus indicates just how prescient urban design can be.

In 1975, New York came close to defaulting on its debt. The infamous headline from the crisis read, "Ford to City: Drop Dead," a reference to President Ford's claim that the U.S. government would let New York collapse rather than provide a federal rescue for city finances. Disowned by the president, fighting with the governor, and hemmed in by bankers who said the city was too financially unstable to be allowed to borrow any more money, the municipal government was put under the control of the Municipal Assistance Corporation, or MAC. The bankers who made up the MAC not only had to approve any financial decisions by the city but imposed their own stiff austerity plans, which led to layoffs for sixty-five thousand city workers, wage and benefits cuts, and service reductions for city residents.[58]

New York's 1975 fiscal crisis is often described as the nadir of New York's midcentury decline.[59] But it simultaneously marked a sea change when urban reinvestment began—on elites' terms. Just as the Chase Manhattan Bank's 1960 headquarters drove the first Battery Park City plan, the hypotenuse-topped Citibank Tower figured into the 1975 master plan for Battery Park City. Citibank was instrumental in the fiscal crisis, since it was Citibank's CEO Walter Wriston who led other bankers to deny the city credit, thus triggering the crisis.[60] Yet that same year, Citibank opened the Citibank Tower.

While the Midtown tower's top is recognized for its shape (and is supposedly slanted as a surface for solar panels), the bottom of the tower is nearly as significant.[61] It is one of the first examples of the trend beginning

Citicorp opened their landmark building in 1975, the same year as New York's notorious fiscal crisis. Though Citicorp had actually triggered the crisis by doubting that the city was fiscally sound, they apparently had sufficient confidence in the city long term to develop the tower. The lower levels have a shopping mall that is an early example of filtered public space.

in 1975 of building filtered spaces—shopping malls—in the bottom floors of major office buildings. The Shops at Citigroup thus memorializes the paradox that just as the bank was claiming New York City was too fiscally unstable for its government to continue to function, the bank was bullish enough on the city to make a major investment that dispensed with privatized plans in favor of a new filtered design. The filtered mall reflected confidence that the moment was near when upper-income professionals could be drawn back to the city, if provided with filtered spaces, to gentrify and recolonize a metropolis that the bank felt had fallen into the hands of municipal unions, low-income recipients of social welfare funds, and the politicians that served those constituencies.

Citibank's simultaneous calculation—that the city as it was was fiscally unacceptable but that Citibank could restructure the local government on the banks' preferred terms to make it attractive for elites once again—is one of the clearest examples of how the design of major projects, particularly their public spaces, not only reflects the zeitgeist of a particular era, but can even be a *leading indicator*. Projects like the Citibank Tower, the Chase headquarters, and most of all Battery Park City itself are the work of people whom John Logan and Harvey Molotch call *structural speculators*.[62] These are large real estate investors who make vast profits not by building for an existing market but by making self-fulfilling predictions about the shape of the city to come. Backed by the coalition that Logan and Molotch call the growth machine (which includes reliably prodevelopment local politicians, newspapers, universities, major businesses, and others who share the class positions, worldview, and priorities of those actors), structural speculators secure state subsidies to build massive projects that change the fundamental uses of a space and its surroundings, potentially delivering "monopolistic" profits to the structural speculators. Because such actors profit to the extent that their projects anticipate future needs rather than merely meet current needs, and because those actors have considerable ability to bring those anticipated needs into being, the projects they build reflect not current social relations so much as the future social relations of a space as elites intend them to be. Consequently, while *gentrification* is a term associated with later periods in American cities, Rouse's filtering festival marketplaces characterized the 1980s, and the first filtered spaces—Faneuil Hall, Citibank, Battery Park City's shopping mall and defensible-space plans—that signified the beginnings of a broad gentrification project took shape by the mid-1970s. Eventually, even more ambitious plans would shape Battery Park City.

Suburban Space Evolves from the 1979 Master Plan

Homeless people began staying in New York's Port Authority bus terminal in the decades after New York's fiscal crisis. Though their presence made the terminal (which accommodated commuter buses from New Jersey and nationwide carriers) less comfortable for travelers, it also provided a place where homeless people could find shelter, sleep, wash and use the restrooms, panhandle, and socialize. One summer in 1992 I returned by bus to New York. Passing through the terminal, I saw police escorting apparently homeless men from the station in what I later deduced was an update of the straitjacket: each man was wrapped in a white bedsheet that was held tight with duct tape wrapped several times around the shoulders, wrists, and ankles. The police could then guide this restrained person wherever they wanted. The person could walk, but only at a shuffle, and could not resist, because without an officer's hand on his shoulder he could crash to the hard floor. I later realized that this action took place immediately before the Democratic National Convention and that the police were probably conducting a sweep to "clean up" Midtown before the arrival of delegates for an event that would put the city in the spotlight. But even after the convention, I never saw large numbers of homeless people in the bus terminal again.

The homeless people in the bus terminal had been removed as part of an extended "sweep" begun under Mayor Ed Koch, continued under Mayor David Dinkins, and accelerated to its most aggressive form under Mayor Rudolph Giuliani. The sweeps reshaped the city by pushing the homeless out of the business and upper-income residential areas of Manhattan. Homeless people were removed not only from the bus terminal but also from Penn Station and the Amtrak train tunnel on the West Side and were beaten up by thugs hired by the business improvement district around Grand Central Station.[63] A new thirty-seven-officer police unit was charged with removing the homeless from places like subway tunnels and public parks.[64] The sweeps rolled from one gentrifying part of the city to another. After sweeps of Washington Square Park, Tompkins Square Park, and Stuyvesant Park, displaced homeless people started sleeping in Union Square, until they were chased out there, too.[65]

These sweeps were undertaken not to help the homeless but to clear them for the benefit of more affluent New Yorkers. As William Bayer, a police commander, explained in 1994, his officers had been told to evict the homeless from parks to make the spaces more pleasant for "law-abiding" people. To the Giuliani administration, the problem was not that homelessness afflicted its citizens, the problem was the homeless citizens them-selves. The police saw the homeless not as a population in need of services but as an "enemy": "We have to cut off the head of the enemy and the en-emy is the homeless," said Captain Bayer.[66]

At the same time the city was kicking the homeless out of Union Square, it proposed cutting funding by $20 million over four years to a "success-ful" program providing single-room-occupancy units to formerly homeless people.[67] Payments to landlords who provided shelter for homeless fami-lies were slashed from $2,300 per person to as little as $1,000 per family.

Among the results of the cuts to housing support were degraded housing options, overcrowding, and, predictably, the relocation of more homeless people to inadequate, "temporary" housing in the poorest and most remote parts of Brooklyn and the Bronx.[68] People were cleared from Manhattan and forced, without services, to find their own way either to low-income neighborhoods far from centers of real estate reinvestment or, given the di-minished services the city now offered, to the edges of the city. New York even moved some homeless people outside the city altogether, to a camp it owned in rural Orange County. County officials there sued for half a mil-lion dollars, contending "that a city-run homeless shelter in their county has become a dumping ground for criminals and the mentally ill."[69]

Homeless service agencies tried to determine where people had been displaced *to*, but they were never able to fully follow up with this transient, difficult-to-reach population. The problem of homelessness had not been solved, but evidence of it had been much reduced in the rapidly gentrifying core of the city.

Sweeps against the homeless were just one element of a broad effort to reshape the city. Rather than cloistering the affluent in filtered spaces, city officials, business improvement districts, real estate developers, and corpo-rate investors sought to remove poor and working-class people, the home-less, and people of color from large swaths of the city that were attractive to real estate investors and their gentrifying customers. The homeless sweeps were just one part of the strategy, along with aggressive police harassment of black and Latino men, reductions in mixed-income and affordable hous-ing options in gentrifying areas, and government support for corporate and luxury development projects (as opposed to support for middle-income, working-class, or low-income housing, for which there was greater demand, or jobs for working-class New Yorkers).

I call this approach the suburban strategy. The suburban strategy within New York applied several of the elements that had made many suburbs bastions of privilege and inequality. First, it replaced the mechanisms of filtering with a plan to separate rich and poor, advantaged and disadvantaged, by vast distances not easily bridged by low-income people. Second, it endorsed state subsidies for market-rate development but refused to develop comparable projects to establish new housing or work opportunities for poor or working-class people.[70] The product of the suburban strategy is *suburban space*, a bifurcated urban reality of affluent spaces that seem less filtered, more open, green, and inviting, and outer-borough spaces where police aggressively surveil people of color with inadequate regard for civil rights, zones to which the state shuttles the homeless. Suburban space is a larger-scale process than initial processes of gentrification. As Sharon Zukin notes, earlier gentrifiers' actions were initially "limited to the small scale of individual neighborhoods and blocks" but were encouraged by journalists' accounts, politicians' ambition, and developers' objectives to create what Zukin called a corporate city, with wide swaths of the urban landscape reconfigured for the benefit of corporate capitalism, real estate development, and chain retail activities.[71]

The suburban strategy proved to be a critical context for the ultimate development of Battery Park City. And Battery Park City proved to be one of the best vantage points from which to see suburban space take shape. It could be that the master plan that Battery Park City adopted in 1979 (and eventually built) was the earliest articulation of the suburban strategy. Certainly, the plan showed a willingness on the part of the BPCA to take a more open, less clearly defended position toward the city. Battery Park City could not become a suburban space without the citywide reorganization reflected in the mid-nineties homeless sweeps and other efforts. If those had not happened, Battery Park City might still be a filtered space. But the plan set into motion the design and planning objectives that would make Battery Park City the epitome of the suburban space, a neighborhood that at once felt open but remained closed, a model for much of the rest of the affluent quarters of the global city in the twenty-first century.

By 1979, the Authority recognized that it needed a more workable development plan that could get construction started. It hired Alexander Cooper Associates in June 1979 to assess the program for Battery Park City, test alternative programs, and design a master plan. It gave the firm four months to do so. Stanley Cooper and Stanton Eckstut's report, published that October, became the master plan that has been closely followed ever since. Searching for a plan that would actually result in more being built, it broke with many of the stylistic aspirations of earlier plans.

For the residential or commercial components of the project to succeed, the Authority needed a plan that would attract developers and promise to attract prospective residents, neither of which it had been able to do for over a decade and a half. During that time, almost nothing had been built, and private developers had shown a consistent lack of interest in building on the site. Battery Park City had so far been a failure, and the Authority, whose entire economic activity up to that point had consisted of using the remainder of its bond money to make interest payments on those bonds, feared being unable to pay its creditors. It needed renewed interest in the site.

But Battery Park City was not, by prevailing real estate measures, a promising site. It was more than a five-minute walk from the nearest subway station. (The 1979 plan tried to deal with that inconvenient fact by misrepresenting the distance to the nearest subway stations.) The Washington Market Urban Renewal Area just across West Street, north of the World Trade Center, was better served by mass transit and was also slated to include a mix of residential and commercial properties.

The Cooper and Eckstut plan was not being developed in the 1990s, when almost any part of Manhattan seemed guaranteed to attract prospective residential tenants ready to pay high rents. In 1979, Independence Plaza, the first element of the adjacent Washington Market Urban Renewal area, had been completed, with subsidized apartments intended for upper-middle-income tenants. It had not been rented out, and eventually the city offered additional subsidies to further lower the rents. Thus the most recent experience in that part of Lower Manhattan indicated that upper- and upper-middle-income people were not drawn to the area. The BPCA and their planners needed to do their best to attract the right kind of people. The text of the plans convincingly demonstrates that the planners dispensed with the preoccupation of keeping people out, in favor of the pressing requirement to find ways of drawing people in. For the first time, security was not the preeminent concern.

The plan's "eight organizing principles" operationalized efforts to attract people. While these kinds of bullet points are sometimes no more than window dressing, in this case they identified some of the primary objectives. The first principle was that Battery Park City "should not be a self-contained new-town-in-town, but a part of Lower Manhattan," criticizing the earlier 1969 plan because it "follows this [self-contained] approach and lays out a series of superblocks with very little relation to the rest of Lower Manhattan. This arrangement contributes to a sense of separateness, *which is considered to be neither a desirable nor a realistic strategy.*"[72] The second

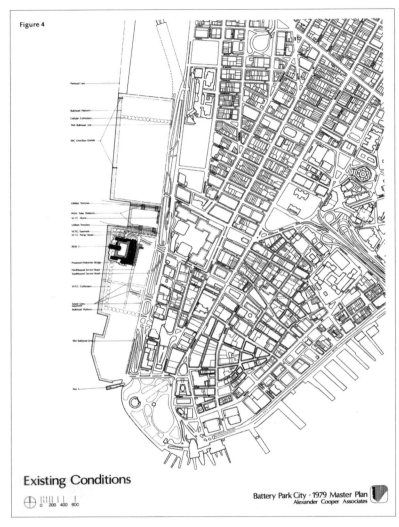

Figure 4

Portcoal Line

Bulkhead Platform
Cellular Cofferdam
1941 Bulkhead Line

BPC Overflow Outfalls

Utilities Trenches
PATH Tube Platforms
W.T.C. Plaza
Utilities Trenches
W.T.C. Basement
W.T.C. Pump House

POD 3

Proposed Pedestrian Bridge
Northbound Service Road
Southbound Service Road

W.T.C. Cofferdam

Sewer Lines
Easement
Bulkhead Platform

1941 Bulkhead Line

Pier A

Existing Conditions

0 200 400 600

Battery Park City · 1979 Master Plan
Alexander Cooper Associates

Battery Park City's final plan, the 1979 design by Cooper and Eckstut, had to contend with a range of preexisting elements. The location of the World Trade Center across from the narrowest part of Battery Park City limited its most valuable real estate. The north end faced vacant lots from an unfinished urban renewal project. The North Cove Marina was already in place to protect the PATH train tunnels. Cooper and Eckstut's street plan followed the sewer lines, and Rector Park covered a sewer intersection that could not be built over. Spatially, there is no such thing as a tabula rasa. (Alexander Cooper Associates, *Battery Park City Draft Summary Report and 1979 Master Plan*, October 1979, fig. 4, p. 40)

principle argued that the neighborhood should not be a self-contained citadel because it would benefit from access to surrounding resources. "The
lessons learned from similar projects elsewhere suggest that Battery Park
City should not turn its back on Lower Manhattan but instead respond to
and build upon the strength and character of the adjacent neighborhoods.
Lower Manhattan's assets are its office inventory, its subway services and its
community facilities and services." The third principle was that "the layout
and orientation of Battery Park City should be an extension of Lower Manhattan's system of streets and blocks" in order to "*overcome a potential sense
of isolation* from the upland area. Utilizing a block system of development
can set a structure for Battery Park City that will easily integrate its building
forms with the adjacent area's existing development. . . . The financial core
will be able to expand more readily onto the project's reservoir of vacant
land. Residents of the project will be better able to support Lower Manhattan's growing range of shops and services. The waterfront amenities at Battery Park City will be more accessible to the employee and resident population of the entire area south of Canal Street."[73] The Cooper and Eckstut
plan exhibited impressive foresight in imagining Battery Park City as part
of a larger, redeveloped Lower Manhattan of improved services, expanded
shopping, and refurbished amenities. Though the area would not take on
these qualities for some time, it accurately described Lower Manhattan in
the period of suburban spaces.

The 1979 plan embraced many of the paradigm shifts of suburban space:
that a space could be more open but no more accessible; that it should
look like the rest of the city but not be intended for use or habitation by
most of the city; and that, in the age of "trickle-down" elitism, the state had
strong obligations to aid high-end development but no evident responsibility for its needier residents. In 1979, however, Battery Park City was not a
suburban space. For many years afterwards, it remained filtered. Even after
several new buildings had been constructed and the waterfront park was
taking shape, it was far less accessible than it is today. In the 1980s, guards
eyed visitors from booths at the entrances to the parks. After hours, guards
sometimes restricted entrance to residents only. The parks closed at certain
hours. Undeveloped lots created a broad, filtering space between the rest of
the city and the new development of Battery Park City. But as Battery Park
City was more completely built out, and as the city engaged in the policing
programs of the suburban strategy, Battery Park City's local barriers began
to fall. Residents remained aware of a physical separation, by West Street
most of all but by other barriers as well. But over time, Battery Park City

came to resemble less a privatized gated community or filtered shopping mall than the wealthy quarters of a suburbanized, imperial global city.[74]

The neighborhood built following Cooper and Eckstut's design was a rare achievement in large-scale projects. Critics celebrated the fact that it felt like a neighborhood. Most such "megaprojects" were notoriously sterile, consisting of vast, anonymous space, huge empty plazas, lonely elevated walkways, and undifferentiated buildings. (The World Trade Center was a faithful representation of that tendency.) To avoid that trap, Battery Park City applied the growing fashion among urban planners of selectively applying the recommendations of the urban critic Jane Jacobs. In *The Death and Life of Great American Cities*, Jacobs challenged modern planning conventions, as represented by Robert Moses.[75] According to her, communities, to be lively and livable, had to have certain features: small blocks to allow easy pedestrian travel, mixed commercial and residential uses to create a diverse street life, "eyes on the street" as an informal way to maintain safety. She cherished older buildings over modernist high-rises both because they fit these requirements and because the building of high-rises tore away and discarded the intricate urban fabric of buildings stitched together over centuries.

Planners did not adopt all of Jacobs's recommendations. She argued that mixed-income diversity was indispensable for a successful neighborhood, but most new places that planners designed were for more homogenous, upper-income groups. She explicitly valued older buildings in the belief that they could provide lower-cost rents (for businesses and residents) than neighboring new structures, thus providing the economic means to maintain such class diversity. In place of such actual rich architectural history, planners in projects like Battery Park City did the next best thing, requiring that buildings *look* like old New York as soon as they were built. Authority building guidelines required buildings to have red-brick facades, avoid the use of mirrored glass and chrome, and install streetlights that were based on a nineteenth-century design. Cooper and Eckstut's original rough drawings of the buildings that would occupy Battery Park City show 1970s-style modernist high-rises and shopping mall interiors. But requirements issued later by the Authority have always required Battery Park City buildings to take a historicist approach to architecture.

Battery Park City thus represents a robust, if inequitable, synthesis of the otherwise incommensurable traditions of Jane Jacobs and Robert Moses. (Jacobs actually spent many years as a well-known community activist opposing several of Moses's Downtown megaprojects.)[76] For this remark-

Images from the 1979 master plan by Cooper and Eckstut outlined where buildings would eventually be built but featured a more modernist look than the historicist style with which the neighborhood came to be so strongly identified. (Alexander Cooper Associates, *Battery Park City Draft Summary Report and 1979 Master Plan*, October 1979, fig. 17, p. 75)

able synthesis, Battery Park City preserved the Moses megaproject but dramatically reduced the public purpose. Meanwhile, planners for the BPCA embraced the letter of Jacobs's recommendations, for short blocks, mixed uses, and old-looking buildings, while dispensing with the actual *purpose* of those requirements, a diverse, mixed-income community. Battery Park City resolved the contradictions of the previous era, creating a useful model for future city building.

Like the models of earlier eras, this new model had its financial landmark. The 1960 Chase Manhattan Bank initialized plans for Battery Park City, seeking to counter the urban flight and disinvestment that had characterized the privatized period. The 1975 Citibank building epitomized the selective return to the city through filtered spaces that had premiered in unrealized plans for Battery Park City. The construction of Goldman Sachs' Group World Headquarters in Battery Park City in 2010 completed the development's suburban period.

Built with $1.65 billion in tax-free federal Liberty Bonds meant to support businesses after September 11, the Goldman building may have been the capstone to the suburban era. Certainly it included elements of the suburban strategy: rather than being sequestered behind privatized space or shopping mall filtering, the lobby was highly visible behind a block-long plate-glass window. Rich ornamental plantings created a green space in front of the building, and the tower itself was green, receiving a gold rating from the LEEDS environmental ratings program.[77]

But because such designs tend to be leading indicators, it may be that the building reflected not the high point of the suburban strategy but, in the midst of the restructuring wrought by the 2008 global economic crisis, a new approach to the control of space. Goldman was certainly at the center of the economic crisis: as the building opened, the public was learning of the firm's involvement in the speculation that had inflated and burst the housing bubble, in the shady debt instruments that had wiped out several large financial firms, and in the misleading accounting that had concealed the extent of Greece's financial crisis, which threatened the rest of Europe.[78] The Goldman Sachs building differed from most of Battery Park City in several ways. First, unlike many office towers in Battery Park City that were connected to the Winter Garden Mall, it contained no public space. Second, an immense mural by the artist Julie Mehretu was highly visible through the full-length glass window to anyone passing by in the traffic on West Street. This diverged from the suburban strategy: it was not open to the gentry the way suburban spaces typically were, since they could not enter the lobby to view the work, which had been reviewed and discussed in publications aimed at the class for whom the suburban space was designed.[79] In addition, the lobby was paradoxically even more visible than suburban spaces, except when the entrance was blocked by crowds of news camera crews covering financial scandals and police and private security guards deployed to protect the building. In this way, though it is too early to know, the Goldman Sachs building may have marked the end of the suburban strategy and the opening of an era that was simultaneously more defiantly visible to the public and more challenged *by* the public than the corporate spaces of the suburban strategy.

Citadel Reality, Ghetto Dreams

Given the beauty, convenience, and quality of Battery Park City at a time of persisting inadequacies and growing inequalities in the city's neighborhoods, it is understandable that critics have assailed the publicly subsidized luxury of Battery Park City for being exclusive. Critics' frustration at tremendous lost opportunities is real: affordable housing that could have been built was not, a community that could have been integrated was instead tightly segregated by race and class. But it is difficult to be angry at an abstract constellation of actors and structures spread over time and space. It is far easier to walk through Battery Park City and dismiss the neighborhood as a gated community designed to keep out the public than to condemn thirty years of backlash policies against subsidized housing at the federal and state level, several mayoral administrations' lip service to eliminating slums, a collection of financial industry elites more interested in making a business district convenient for themselves than for most of the people who worked there, and progressive activists' failure to mobilize sufficiently to win affordable housing for all. But the visceral satisfaction of such a condemnation is less valuable than the advantages to be gained in counteracting the actual structural forces at work. Battery Park City is a good symbol of problems that trouble progressives and New Yorkers generally, and a disturbing reflection of the assumptions of the suburban strategy that motivate public development projects today. But few of the problems it represents are Battery Park City's alone.

Battery Park City's history demonstrates why the neighborhood has ultimately proved paradoxically exclusive *and* vibrant. The financial and political elites who conceived of Battery Park City and shepherded it through decades of political opposition consistently sought to build an exclusive citadel. But elites' changing understanding of how they could use the city created a much more lively community than a whole generation of master plans had initially envisioned.

Once nearly all of Manhattan had been claimed for high-rent residences and businesses, walls around Battery Park City proved unnecessary, as the command-and-control functions ascribed to the citadel came to dominate the landscape beyond Battery Park City's borders as well. The citadel and the ghetto correctly described an economically and racially segregated city under global capitalism. As the next chapter shows, however, these polarized communities have not grown up side by side but are spatially distant. Reducing class mixing and introducing large swaths of economic homogeneity in the core of New York City are components of the suburban strategy that have informed development and planning in New York in the last two decades.

2

Real Privilege and False Charity

WHEN ELITES PLANNING Battery Park City successfully excluded affordable housing from the neighborhood, demands for such housing did not disappear. Eventually, excess revenue from Battery Park City was slated to build and renovate a considerable quantity of affordable housing—but it was to be far from Battery Park City itself. This chapter examines the real privileges that accrued to people who lived in Battery Park City because of the special arrangements made by the city and state for this gilded citadel, and sets them against the false charity of the affordable housing plan.

Battery Park City had an unacknowledged sibling. The New Settlement Apartments, in the Mt. Eden section of the Bronx, were built in 1991 with surplus funds from the operation of Battery Park City. Both Battery Park City and New Settlement Apartments were composed of about fifteen buildings.[1] Both complexes were well built and exceptionally well maintained, with a higher density of community organizations and services than surrounding neighborhoods. But in New Settlement rents were moderated, and 30 percent of households had been formerly homeless. The two neighborhoods were separated at birth by race, class, and an hourlong subway ride. But fittingly for a port city, they were connected by the water.

At the North Marina in Battery Park City, members of the Manhattan Yacht Club maneuvered some of the dozen identical twenty-four-foot sailboats belonging to the club out of the tight confines of the marina's protective breakwater and headed out into the Hudson River. The waves were surprisingly large on the Hudson, and the current was often strong. From the shore I had often watched as the boats moved expertly through New York Harbor between priceless views of the Statue of Liberty and the towers of Battery Park City. Membership at the yacht club cost a few thousand dollars per year, more for the handful of members who owned their own sailboat. While members were not necessarily connected to Battery Park City, the club's commodore, Michael Fortenbaugh, had become a well-known figure in the Battery Park City community as active residents asked the club for support on local matters, and the club's efforts to reopen and expand after September 11 were discussed at Community Board meetings. The marina, as home for the yacht club, provided recreational boaters in Manhattan a rare opportunity to sail close to home.

In the Bronx, another group was launching their boats. For almost a decade, Bronx high school students in the after-school program Rocking the Boat would spend a year building large wooden rowboats based on traditional designs. The group's first workshop was in the New Settlement Apartments, and many of its members were Latino and African American students from New Settlement and the surrounding Mt. Eden area. I attended the annual June 2006 barbeque at Clason Point Park where the newest boat was launched. The park was newly renovated, an open, grassy, still rough-at-the-edges space with a ramp down to the water of the East River. Barbara, a social worker with Rocking the Boat who worked with students on academics, social and emotional issues, and plans for college or vocational education after high school, invited me out on the boat. Students gave us life jackets, and we managed to keep our feet dry while climbing into the *High Tide*. Two Rocking the Boat members, including Eddie, one of the senior apprentices, rowed us out. They worked the boat into the wind, maneuvered it out beyond the shore, and rowed around the rusting hull of an abandoned, half-sunken ship. I asked Eddie, who had worked in the boat program for several years, and lived in New Settlement Apartments, what he thought of the boat program. "It's great," he said with enthusiasm. After working with Rocking the Boat, he wanted to be a boat builder —"or," he corrected himself, "a carpenter." It sounded as though someone had suggested that there were more jobs for carpenters than for boat builders, and while he had tried to internalize that message, it was hard because it was boat building that he had come to love.

Given Battery Park City's affluence and New Settlement Apartments' poverty, one might wonder why there wasn't a partnership between institutions like the Yacht Club and Rocking the Boat. In fact, Rocking the Boat's founder Adam Green said Commodore Fortenbaugh had made such a suggestion. Packing up at the end of the day of boat launches and barbequing, Adam explained. Fortenbaugh had said to him, "What I want is for your guys to build a boat on a barge there [in the North Marina] so the whole financial community is looking down on you." Green paused a split second at the double entendre of the financial community "looking down on" the kids.

"We don't really work with that community," Green deadpanned about the Financial District. But, insisted Fortenbaugh, it would be an opportunity for Green to bring Bronx kids down to the Financial District.

Green bridled. "Our kids' parents already work down there for you guys. So why would we want to send them down there?" he asked. While to Fortenbaugh Wall Street was a rich resource he was being generous in trying to share with less fortunate kids, to Green (caught, like several longtime Bronx activists, between the class conferred by their college degree and the class of the Bronx communities with which they were strongly allied) Wall Street represented not opportunities provided but dreams denied—the power and privilege accumulated by extracting labor from working people like his students' parents. Green and others active around New Settlement bristled at what the social activist and educator Paulo Freire had defined as "false charity," or largesse that still maintained the disadvantaged or subservient position of the recipients. "In order to have the continued opportunity to express their 'generosity' the oppressors must perpetuate injustice as well," Freire wrote. In contrast, "True generosity lies in striving so that these hands—whether of individuals or entire peoples—need be extended less and less in supplication, so that more and more they become human hands which work and, working, transform the world."[2] Battery Park City's abundance was real privilege. Liberal offers to share that privilege were false charity, and organizers in the Bronx saw them that way.

As people around the New Settlement Apartments like Adam Green made clear, the charity pointed to with pride by the Battery Park City Authority (BPCA) and residents alike rang hollow to those outside the citadel's walls. Detailing Battery Park City's privilege in this chapter serves several functions. First, it continues the previous chapter's examination of how privilege was constructed in Battery Park City: the last chapter considered the evolution of design exclusion, while this chapter looks at the programmatic and organizational choices that produced financial benefits for residents in Battery Park City. Second, it describes the financial, programmatic, and cultural advantages that Battery Park City residents have enjoyed, and that, as subsequent chapters show, they have worked so hard to preserve. Third, examining the common justifications for these privileges reveals the seemingly commonsense rationalizations for privilege that are ingrained in the suburban strategy today.

Installing Privilege in Battery Park City

Battery Park City was the beneficiary of several subsidies. The most critical framing of those subsidies would say that Battery Park City didn't pay its fair share of taxes, refused to build housing that most people could afford, constructed itself as a racially segregated community, subsidized the rents of its already wealthy residents, and kept out the homeless by privatizing public space. Each allegation is true. Defenders of Battery Park City have long provided narratives to justify these actions. As the sociologist Charles Tilly pointed out in his book *Why?* such stories and formulaic justifications are most successful when used by people in positions of greater power to explain things to those in positions of less power. The stories, therefore, should not be accepted as justification but challenged for the inequality they shield. This section examines the most significant privileges Battery Park City enjoys, and it considers the critiques of, and justifications for, those privileges.

The Battery Park City Authority and Payments in Lieu of Taxes

Battery Park City's first subsidy is the neighborhoods' unusual ability to use its tax revenues for its community alone without being forced to pay for citywide services. Battery Park City is owned and operated by the BPCA. One of 169 state authorities (including the Metropolitan Transit Authority and the Port Authority of New York and New Jersey, as well as forgotten entities like the Overcoat Development Corporation), the tax-exempt BPCA was established to oversee the financing, design, and construction of Battery Park City.[3] Its directors are appointed by the governor. The BPCA, not the city or state, is thus directly responsible for decisions about neighborhood development, the presence of commercial, retail, residential, or affordable residential buildings, the design and maintenance of parks, and even some quality-of-life policing. Many community concerns are therefore directed at the Authority, which residents feel has been more or less responsive to residents depending on who is in charge of the organization.

While other city residences are assessed New York City property taxes, the buildings in Battery Park City make "Payments in Lieu of Taxes," or PILOT, to the Authority. The Authority uses as much of this money as it wants to operate and enrich its community, then turns any remainder over to the city.

The financial benefits of such an arrangement are enormous. Unless the Authority decides to turn over surplus PILOT money, Battery Park City residents do not pay property taxes for all the city services, like education, health care, police and fire protection, sanitation, and snow removal, that they and other city residents use. In a ten-year period, the Authority took in $115 million in revenues but turned over only 31 percent to the city. According to analysis by the group City Project, over a decade those payments of surpluses to the city amounted to only 19 percent of the property taxes the community would normally owe.[4] Any neighborhood would be improved if it could direct its collective tax money to its own needs and still benefit from the city services paid for by others. Indeed, the Authority spent twenty-five times per acre what the city spent on other city parks, not surprisingly creating a string of emerald parks around the neighborhood that are the envy of anywhere else in the city.[5]

The justification for such a financial arrangement is familiar to anyone who studies large-scale downtown redevelopment projects. PILOT programs, like "tax increment financing," are a way to publicly finance projects without giving the appearance to the public that general funds are being spent on a project with primarily private benefits. Boosters of such projects argue that the final product attracts businesses, creates jobs, and revitalizes the city in a way that benefits all. Such claims are typically overblown (as they are in Battery Park City, where the businesses that moved to the World Financial Center were already well-established New York institutions, not new businesses). Critics of such tax-redirection schemes also ask whether the money could have been better used, been directed toward a broader strata of the city's workforce, or included jobs or housing for more low-income workers. When such questions are raised, as the next section details, conventional wisdom suggests there is no alternative.

Opportunity Costs and Affordable Housing

An important subsidy for upper-income Battery Park City residents came about in the creation of an upper-income residential community itself. Many have argued that Battery Park City, as a state project, should have been a low-income or mixed-income community. Defenders of Battery Park City's elite profile dismiss such ideas as unrealistic: there was no money available to build affordable housing. But a closer examination shows that in fact significant sacrifices were required to construct an *upper-income* community.

Supporters of Battery Park City's luxury residential profile, such as the planner and researcher David Gordon, suggested that a luxury profile was necessary for the vitality of Lower Manhattan. An expensive neighborhood made it convenient for executives and kept financial service firms from leaving Lower Manhattan.[6] But companies also choose their location for access to ready pools of employees; proponents do not make clear why the city and state should have subsidized housing only for Wall Street executives and not for the rest of the industry's workforce. This lapse was all the more striking given the demand for housing in Lower Manhattan: in 1967 and 1971, a Downtown business group's survey of the area found "demand for mid-income housing strong, demand for low-cost housing essentially unlimited, and demand for high-income housing quite limited, consisting almost entirely of a few people who would be interested in moving to the area from other parts of Manhattan."[7] Thus, Maynard Robison wrote in his research on the early planning of Battery Park City, the project's promoters and developers set up an unnecessary development hurdle: "There was no real demand for expensive housing in the area."[8] Contrary to claims by proponents like Gordon, though luxury housing is often called "market rate" in New York, ironically the decision to build it was not market driven, since it was the only use of the land for which there was no ready market.

Though the Authority did not plan to build significant affordable housing *in* Battery Park City, it had planned to provide money to build affordable housing far away from Battery Park City, in predominantly black and Latino neighborhoods. Meyer "Sandy" Frucher was president and chief executive officer of BPCA from 1984 to 1988, the period during which residential construction began in earnest in Battery Park City. I interviewed Frucher in a classic Manhattan diner, the kind I had already found that a senior generation of New York's city builders preferred for interviews. Perhaps they liked such places because they could combine coffee and an interview, perhaps they preferred the unpretentiousness of such a public setting, or perhaps they didn't want interviewers in their offices or homes. Over breakfast, Frucher made it clear that he had only ever intended to build upper-income housing in Battery Park City. But he was proud of an agreement he had negotiated with his friend Governor Mario Cuomo to use surplus revenues from the luxury housing in Battery Park City to build or renovate lower-cost housing in Harlem and the Bronx:

> He [Cuomo] thought that Battery Park City lacked soul. It might be a good development, but what was the state doing involved in top-of-the-line commercial and residential? He didn't like the notion of just building upscale. So the bottom line of our conversation was, he effectively said, "Get it built, give it a soul." . . . So what we did is we came up with

a multiphase strategy. Part A was to make the public spaces terrific, and to develop public facilities at Battery Park City that became destinations for the whole city. Part B was to generate funds from Battery Park City to create the housing program which would enable . . . low and affordable housing would be able to be built outside of Battery Park City out of the proceeds of Battery Park City. So that became a dual strategy.

Frucher called the plan to direct Battery Park City funds to offsite affordable housing "the best thing I have ever done—professionally," because it gave Battery Park City the public purpose Cuomo had demanded.

Critics pointed out that the scheme to build affordable housing away from Battery Park City segregated the poor and their subsidized housing far from Battery Park City. Housing justice and integration would require that affordable housing be built in Battery Park City itself. Gordon chastised such critics, arguing that they ignored how expensive it would be to use high-rent land for low-rent apartments, and that the money it would take to pay for such land should instead pay for more units on cheaper real estate. "These critics seemed to demand a symbolic victory—fewer affordable housing units but a mixed income neighborhood."[9] But it has been the funding for off-site housing that has been symbolic. While the BPCA, through the program Frucher developed, promised $1 billion for affordable housing elsewhere in the city, only a handful of units were renovated, and no more than $143 million was actually used for housing, for New Settlement Apartments and a series of renovations in Harlem and the Bronx. (Though more than this amount has been delivered to the city, it has been absorbed into the general fund and not used for housing as promised, even when the city recorded record surpluses during the Giuliani administration. This violated the memorandum of understanding between the BPCA and the city.)[10] Nor, contrary to Gordon's suggestion, would affordable housing have been more expensive to construct in Battery Park City than in lower-income areas like the Bronx or Harlem. Sandy Frucher said that land costs were the same in Battery Park City as in low-income neighborhoods. Contradicting scholars like Gordon who thought building affordable housing in Battery Park City would have been more expensive, Frucher said, "There's no question the land costs, the land costs in ghetto areas at that time were probably cheap too, it wasn't a question of the land cost." Particularly because Battery Park City had been created as landfill, the price of the land was not high. But Frucher and others in a position to decide the fate of that land did not want to build low-income housing there. Even if surplus Battery Park City funds were to be used for affordable housing (and Frucher said he believed they should be, in keeping with the spirit of a public project), they would be used to build housing elsewhere.

Frucher articulated the rationale for building affordable housing outside Battery Park City differently than did Gordon. As he described it, low-cost housing required two subsidies: public funds to build it, and subsidies to cover operation and maintenance costs, which Frucher argued could not be covered by tenants' rents alone. To build an economically (and racially) integrated community in Battery Park City would have additionally required a third subsidy, which Frucher described as "an opportunity cost subsidy, which is, if I give up the site that's going to generate a heck of a lot of revenue, and build low-income housing, then I don't have the revenue to do any of the subsidies. So by maintaining my revenue stream and my tax base [by building only luxury units in Battery Park City], I can build a hell of a lot more because I only have two subsidies, instead of two subsidies plus giving up the revenue which effectively becomes three subsidies." Frucher was committed to building affordable housing, but he suggested that foregoing the maximum rent on Battery Park City's land was one subsidy the Authority could not bear.

Yet by building a luxury residential neighborhood in Battery Park City, the Authority had paid just such an "opportunity cost." Some of the highest land rents in the United States are generated by business uses of Lower Manhattan and are above those even of luxury residential rates. The Authority was willing to suffer the "opportunity costs"—the loss in optimal revenue—from using most of its land for residential instead of business uses. But it fought resolutely against paying the "opportunity costs" entailed in making that housing affordable. Thus, while Frucher explicitly contradicted Gordon's conventional excuse for the absence of affordable housing in Battery Park City—that it would have been too expensive—he simultaneously demonstrated the limits of market economics as an explanation for the shape of the luxury community. Opportunity costs were paid to create a luxury community. But building an integrated community was considered excessive.

The difference in cost between integrated and segregated communities was less than some had argued. Ultimately the real opportunity—to create affordable housing and an integrated community—was missed. Battery Park City would be segregated.

In the end, the city did not use most of the money it received from Battery Park City for housing. Knowing that the money was being misappropriated, even when the city had a surplus, the Authority could just as well have used future surpluses to build affordable housing, or even mixed-income projects, on their remaining sites. They also could have leased land for mixed-income projects, suffering nothing but the hypothetical loss of higher rents for the land. Thus while government sources for housing subsidies were long gone, the Authority itself had the means to make Battery Park City more diverse and inclusive and to serve a broader cross section of taxpayers than it did. As Maynard Robison found, conversations with those intimately involved in planning Battery Park City made clear that it lacked affordable housing not because the funds were unavailable but because decision makers did not want affordable housing there.

"Visual Diversity" as Racial Privilege

Elites' steadfast resistance to economically integrating Battery Park City revealed a fundamental resistance to racial integration as well. While activists who work for open, nondiscriminatory housing and unsegregated neighborhoods see integration as a public good, surveys find that whites often endorse integration in theory but oppose it when it is proposed for their actual neighborhood.[11] Examination of the planning of Battery Park City and the proposal to build affordable housing elsewhere indicates that elites (both the planners and the wealthy residents of Battery Park City) sought enough residential diversity to exonerate themselves of potential allegations of racism but ultimately perceived integration as a *burden*, and one to be borne by others.

Frucher reflected this constrained view of integration. In making the case to Governor Cuomo that Battery Park City itself should not be integrated and that low-income housing belonged far away, Frucher recalled asking, "*At what cost integration?* I mean, does every block have to be integrated, does every unit have to be integrated, every building have to be integrated? Is it a virtue in a city to have Park Avenues, and Battery Park Cities that generate sufficient revenue that allows you to go out and then build? You can't do it without taxes, you need a tax base." Yet whether Park Avenue was integrated or not, tax revenues would be collected, from rich and poor households alike no matter where they lived, and could be used to build future integrated communities. In fact, building luxury homes in Battery Park City, which paid no property taxes, made such funds less available than they would have been otherwise. Decision makers like Frucher argued against integration but did so in a form that allowed them to support integration in theory but never implement it in practice.

This is not a critique of Frucher's personal positions. In fact, Frucher's biography complicates the story because he has been actively involved throughout his life in efforts for racial progress and programs for the poor. A professor who knew him as a college student recalled his involvement in civil rights issues. Frucher joined the board of the nonprofit that oversaw New Settlement Apartments. He worked in support of that low-income project and facilitated construction of a new public school in New Settlement's neighborhood. While at the BPCA Frucher set up a graduate school scholarship program to bring more students of color into real estate development. Among those elites best positioned to reduce inequality by reshaping the landscape, however, such commitments exist on an unequal standing with fealty to economic power. Unlike low-income community activists, Frucher saw no contradiction in building a segregated, upper-income neighborhood and a high-quality, low-income neighborhood. He had opposed the status quo by developing affirmative action programs but served the status quo as CEO of the Philadelphia Stock Exchange. Elites who built Battery Park City voiced *theoretical* support for integration, improved housing for the poor, and contributions to the public coffers, consistent with the norms of contemporary discourse.[12] But practical support dried up when it challenged their image of the appropriate kinds of spaces for elites, as it did in the planning of Battery Park City.

Such an apparent contradiction—that the same person would support liberal ameliorative programs and work on behalf of other projects that contravened those efforts—is not one individual's idiosyncrasy but is representative of a larger class of elite liberal. The paradox is well captured in David Halberstam's phrase the "best and the brightest." And while Halberstam used it to describe ambitious officials who during the Vietnam War pursued "brilliant policies that defied common sense," the term and its irony equally describe mainstream urban liberals.[13] Mayor John Lindsay and Governor Nelson Rockefeller, after all, are recalled as social liberals, but they endorsed an elite profile for Battery Park City. Mayor Lindsay is recalled as an elite "good government" "moralist," but he did nothing to stop the plague of fires in the 1970s and 1980s that resulted when the city cut back fire services to poor neighborhoods in a policy of "planned shrinkage" that burned thousands of predominantly poor, black, and Latino New Yorkers out of their homes.[14] Unlike civil rights activists, liberal politicians responding to demands for equality or integration were characterized by split responses that created some progressive programs but maintained other segregated policies. Residents of Battery Park City, many of whom were in the same class position, exhibited a similar selective embrace of values such as integration.

Battery Park City's selective embrace of diversity reflected elites' attitudes. By one measure, after all, it was not highly segregated: the neighborhood was almost 25 percent nonwhite. That figure is low for New York City as a whole but shows more diversity than most neighborhoods of upper-income whites. However, African Americans and Latinos were underrepresented: there were four to six times more Asian residents than black or Latino Battery Park City residents, though the overall city's Asian population is considerably smaller than its black and Latino populations. Some Asian residents have commented on the feeling of being the one group allowed to lend "visual diversity" to the community. Manohar Kanuri, an Indian-born resident, assessed racial diversity in Battery Park City and found it clearly lacking but suspected that the community seemed diverse enough for many of his neighbors:

> That's one of the things I've noticed is [there are] Asians—South Asians like me. But that's as far as it goes . . . But blacks? All through the thirteen years that I've been here I can count the black families on my one hand. That's always been the case. Yeah, I used to have one woman who's a neighbor who was black, but like many lily-white places [black residents in Battery Park City] tend to be African origin, not African American origin . . . There's the honorary white status of East Asians . . . I think there's definitely a lot of people who are attracted to it [Battery Park City] because it's white. I don't discount that at all. And I think there's a big element of the faux liberal who appreciates the visual diversity of it . . . Where are the black families? . . . That's unfortunately true of a lot of Manhattan. Few [non-black or non-Latino] people notice that, because there's a lot of people who work in the city during the day, so visually they're satisfied, on the subway, they're rubbing shoulders, so it satisfies the expectation that okay, I'm rubbing shoulders, I'm diverse, and that goes to everyone whose self-conception is liberal. After all you have to confirm for yourself that you're liberal in some way.

Kanuri suggested that "faux liberal" whites were satisfied by a superficial diversity supplied by the presence of Asian neighbors and racially diverse commutes on the subway, and didn't notice that the neighborhood's diversity did not include adequate representation of blacks or Latinos. But he reported that whites in his neighborhood expressed their racial views to him because of his status as an honorary white. Over lunch one day, Kanuri explained that another former neighbor "was quite explicit" about his preference for Battery Park City as a racially white place. He said that in conversations, at least some white neighbors would look around to see who might be in earshot, then describe the neighborhood as a place that was "closer to Wall Street than the Upper East Side, and you don't have any black people around on the streets." I did not hear such comments directly, though there are several reasons I was unlikely to.[15] Researchers have examined in detail how white Americans today avoid talking about race and use a "color-blind" discourse even when discussing racist attitudes, and they have similarly documented that most whites prefer overwhelmingly white neighborhoods. Battery Park City is probably no different from most such places.[16] Thus when Frucher, in asking, "At what cost integration?" assumed that there was a disadvantage to integration, he had accurately taken a sounding of this wealthy white New York neighborhood's attitudes toward integration: segregation had a value, whereas integration carried some kind of cost.

The class structure of U.S. immigration flows and the relatively high proportion of Asian and South Asian immigrants entering the United States with professional preference visas (allowing them to come to the United States because of their skills in a technical field with a shortage of workers in the United States) helped ensure that there would be a meaningful Asian population in a neighborhood like Battery Park City, but in the absence of affordable housing and affirmative marketing campaigns to encourage greater diversity, African Americans and Latinos would be sorely underrepresented. Battery Park City's composition thus appealed both to those Kanuri called "faux liberals" looking for "visual diversity" and to nonliberals who were explicit about their preference not to live near African Americans. The fact that whites in the United States interpret living in highly segregated, even all-white communities as "normal" meant that whites were unlikely to be disturbed by Battery Park City's segregation (which was, after all, less pronounced than in many other places in the United States, and even in New York) and to move into the neighborhood without reservations.

The development of Battery Park City illustrates how relatively non-controversial beliefs about appropriate land use and neighborhood composition produce segregation even in brand-new neighborhoods built by public agencies ostensibly in the public interest. While other authors have made compelling cases for the need for aggressive public policies aimed at desegregation, such as expanded use of local tester organizations, realtor audits to look for discriminatory practices, investigation of discriminatory lending practices, desegregation of public housing through the use of rental vouchers, and facilitated means of bringing housing discrimination cases, this case provides evidence that there are other problem areas to be addressed.[17] At the least, it shows that one component of the suburban strategy's ideology is the abandonment of any public commitment to integration by an influential group of decision makers. (One implication is that while New York has often been celebrated as a city of closely mixed economic groups, that quality could become a relic of the past.) As long as the assumptions about land use and community building that undergirded the programmatic planning of Battery Park City are in place, segregation is likely not only to be preserved where it exists but to be extended into new developments. State development authorities reproduce racial and economic exclusion. Changing those agencies' priorities will be necessary to produce more integrated neighborhoods.

Through its presentation of visual diversity, Battery Park City reflected the tendency to pay lip service to "diversity" (more often than to "integration") in theory even as Battery Park City's planners and many of its residents endorsed a more racially restrictive image of what the neighborhood was. Whether living in a racially segregated neighborhood is actually a privilege or a disability is difficult to assess, notwithstanding the advantages accruing to various types of white neighborhoods. However, the ability to direct state financial, planning, and construction resources to build a neighborhood consistent with one's own preference for a racially segregated neighborhood reflects considerable power on the part of planners and residents alike. From the outset, Battery Park City was accorded benefits few others could obtain. This proved to still be true, though in a very different context, in the years immediately after September 11, when a new, emergency subsidy was introduced in Lower Manhattan.

Rent Subsidies after September 11

After September 11, Battery Park City was the beneficiary of another subsidy, one that was generous and thoughtful but lopsided. The Lower Manhattan Development Corporation, citing concern that there would be weak demand for housing in Lower Manhattan as a result of September 11, announced in 2002 an Individual Assistance Program that would pay residents up to $12,000 per household over two years. The goal of the program was both to ease financial hardships that residents were facing in the wake of the attacks and to entice newcomers to move to apartments that others had left, countering high vacancy rates and a threat to property values and neighborhood vitality.[18]

Residents, who had already begun forming tenants' associations to negotiate rent reductions on their apartments (particularly when the buildings had been uninhabitable for several months), appreciated the program. Some complained that the enticement had brought short-term residents to the area, that some college students took advantage of the $500-per-month rent reduction and packed into apartments. Residents expected that most would move out when the two-year deal expired. There were occasional whispers that the various forms of financial assistance available to residents after September 11 could hinder rather than help people move on with their lives. This concern was built on the model of public discourse that imagines welfare programs to carry risks of "dependency." A relatively small minority of residents therefore thought that some of their neighbors might be rendered dependent on, or enamored of, the generosity of programs like the rent subsidies and other post–September 11 financial assistance. But among most residents the rent subsidy program was popular, and the broader public viewed the subsidies as a wise breakwater against a possible surge in vacancies that could have brought long-term damage to the viability of the then-fragile neighborhood.

The program may well have been a prudent guard against higher rates of departures and an incentive that brought newcomers to revive Lower Manhattan. But eventually people from other neighborhoods voiced frustration. When neighborhoods like the blocks of Mt. Eden near the Cross Bronx Expressway and Grand Concourse that would later be known as the New Settlement Apartments faced even more severe threats from depopulation in the 1970s, the city not only did nothing but cut city services to catalyze abandonment. Photos of the Mt. Eden buildings before they were rehabilitated can be seen in the New Settlement management offices today. They were completely abandoned, with gaping holes for windows. The shell of a junked car was perched in one entry courtyard. In such neighborhoods across the city, the government didn't even provide the resources to board up dangerous abandoned buildings.[19] Unlike Battery Park City, neighborhoods like Mt. Eden, which had housed middle-class, working-class, and poor residents, had been not just ignored in their time of distress but further assaulted. Even necessary programs become privileges when they are inequitably distributed. That Battery Park City and other Lower Manhattan neighborhoods got what they needed in a time of crisis is not blameworthy, except that so many other neighborhoods did not.

Zoning That Privatizes Control of Public Space

Another overlooked aspect of the design of Battery Park City is the distinctive regulations that govern construction there. Because it is a state project on state-owned land (developers hold only a ninety-nine-year lease), it is governed by a set of zoning regulations different from those governing the rest of Manhattan. Thus, while city zoning encourages other high-rise developers to provide public spaces such as plazas, arcades, and parks in exchange for a "bonus" in the permissible size of their building, that "bonus" regulation was not present here. Instead, in Battery Park City developers provided funds for public spaces in a way that made parks more accessible but plazas and malls less so. Buildings' payments in lieu of taxes to the BPCA went to the upkeep and maintenance of the public parks because the Authority wanted to ensure that they could be maintained to a higher standard than would have been possible with only the funding of the city's Parks Department. The effect was to bifurcate the neighborhood's public spaces: outdoor parks were better maintained than most privately managed spaces in the city, but indoor public spaces didn't have the guarantees of public access that bonus spaces elsewhere in the city carried. Thus guards at the Winter Garden Mall selectively and explicitly excluded homeless people, as they could not have done without cause in bonus spaces like the Citicorp Building Mall in Midtown Manhattan. Something as arcane as the zoning code shaped who used and who was excluded from the neighborhood.

Access by homeless people is a sensitive indicator of accessibility: the presence of visibly homeless people typically indicates that anyone can get into a place, even if the places used by the unhoused are not desirable to everyone. Susan Fainstein, in one of the most significant analyses of Battery Park City, noted in passing that Battery Park City was not much used by homeless people. This matched my own observations of Battery Park City, which rarely had homeless people in the neighborhood, compared to other parts of Lower Manhattan, including the adjacent Battery Park. Fainstein concluded, from conversations with the BPCA and homeless service providers, that this was because "physical and psychological barriers keep out such people."[20]

My field observations of Battery Park City public spaces found them almost devoid of homeless users. This is in part because of the legal regime under which it was built. Malls and indoor spaces along nearby Wall Street built under the bonus plaza program were required to be open to the public. As a result, homeless people regularly sat quietly in them, particularly on days when the weather was harsh. But a space like the Winter Garden, though it looked like many indoor malls built under the bonus plaza program, was entirely private space over which the owners had control. I asked a security guard questions about who used the Winter Garden Mall. He explained that he allowed businessmen to eat lunch and mothers and nannies with babies to stay in the glass-enclosed courtyard. But if there was someone sitting or sleeping there who looked homeless, he showed them out, he said, emphasizing the point with a dismissive sweep of his hand. Because of the distinctive governance regime of a city neighborhood that was state controlled and state owned, the public parks could be more lavishly maintained, while other public spaces were simultaneously made less publicly accessible.

The quantifiable privileges in Battery Park City therefore came from a range of specific policies, but they generally shared several features. The state was *able* to provide most of these benefits because it had organized Battery Park City under the auspices of a state authority. More importantly, however, local, state, and federal actors *chose* to provide unusually generous services because, as in the case of racially segregating Battery Park City, the decision by elite decision makers to do so was consistent with their own conception of what was appropriate for residents who were similar to themselves in status and privilege. In an important sense, decision makers like Sandy Frucher were making decisions not for "them" but for "us" — people of a social status similar to their own. Their rationales reflected the ideological justifications of the suburban strategy for the uneven distribution of public goods. Under the suburban strategy, it is unproblematic for the state to provide additional benefits to an already privileged class rather than direct public benefits to members of society more in need of them. Therefore it did not bother government officials that Battery Park City residents would get city services without paying city taxes. It seemed natural to suffer opportunity costs to build the wealthy residential neighborhood, but excessive to accept the opportunity costs required to make that neighborhood economically inclusive. It was important that the neighborhood make appropriate gestures toward liberal notions of diversity and demonstration programs of affordable housing, but decision makers were nonplussed at the suggestion that it would have been better public policy to not make the neighborhood as racially segregated as it was. In all likelihood, officials at the Lower Manhattan Development Corporation accurately reflected the sympathy the country felt toward people who had lived around the World Trade Center when it unveiled the $12,000 rent subsidy program. But while this may have been appropriate, public silence in the face of larger disasters that afflicted poorer urbanites resonated that much more loudly. And while city officials at least periodically enforced the public right to access bonus plazas in other parts of the city, BPCA officials did not feel compelled to guarantee people the same right of access to plaza and mall spaces in Battery Park City buildings.

The record of privileges bestowed on Battery Park City helps explain an important source of residents' fierce defense of their community against perceived intrusions by outsiders. While Battery Park City's physical isolation contributed to a sense of distinction between residents and non-residents, chapter 3 details how that physical reality interacted reciprocally with the social reality of privilege to create a definition of community for residents that was highly reliant on the exclusion of those outside the neighborhood's political and physical boundaries.

Defending Real Privilege with a Display of False Charity: New Settlement Apartments

When Battery Park City's pronounced privileges are criticized, defenders point to the charity that Battery Park City's privileged position has allowed it to bestow on others. In particular, the affordable housing program developed by Sandy Frucher has been held up, by Gordon and others, as a concrete justification for Battery Park City's otherwise lopsided subsidies. Frucher made such an argument to a city councilman who had argued in favor of making Battery Park City an inclusive, mixed-income community: there wasn't enough affordable housing, so if by making Battery Park City entirely a luxury community Frucher could generate a bigger pool of surplus funds to be directed to affordable housing, wasn't that the best route to the greatest good? In Frucher's telling of the story, the contrite councilman was convinced.

The fact that little of that surplus money was ever used to build affordable housing made such a defense moot in this case. But what about the more general justification for the suburban strategy of real estate development: that welfare for the wealthy will trickle down to the rest of us? The rationale for the larger implementation of the suburban strategy in New York, concomitant with the reorganization of the city in the interests of financial capital, is that the financial activity of that sector will benefit the city as a whole. Spending money on elite projects like Battery Park City is justifiable because they will later generate revenues to support programs for those who are economically excluded from the elite projects. New Settlement Apartments, the largest beneficiary of Battery Park City affordable housing funds, is the best place to assess such an argument. In fact, activists in New Settlement were not convinced by Battery Park City's charity: they argued instead that their community's needs should take priority over the luxuries destined for their downriver twin.

Certainly, organizations that had been able to build affordable housing with Battery Park City money appreciated the arrangement. Carol Lamberg directed the Settlement Housing Fund, the nonprofit that oversaw rehabilitation of the Bronx apartment buildings that would become the New Settlement Apartments. In an interview at the organization's Manhattan offices, Lamberg described her work with Sandy Frucher to secure state approval of the plan to build affordable housing with Battery Park City money. Battery Park City's funding of the renovation, which freed the housing group from paying off a loan for the project, meant that the group had the freedom, and relative financial independence, to put more effort into organizing the community. As another staff member at Settlement Housing Fund explained, the additional money meant that the complex could support resident organizing without concern for attracting—or offending—external funders. Thus, when residents wanted to organize to demand improvements in the local schools, Settlement Housing Fund didn't have to worry that the group's militancy would offend charitable foundations or other institutional funders. The Fund could provide support for the parents' organization itself. Thus the community benefited directly from its unique funding arrangement with Battery Park City. Other connections were maintained for a time between the two separated-at-birth communities. Lamberg recalled that Tessa Huxley, the head of the Battery Park City Parks Conservancy, came up to New Settlement Apartments to contribute her considerable expertise in sustainable urban gardening. An enthusiastic BPCA board member was so taken by a tour of the New Settlement Apartments that he funded a brochure that celebrated the connection between the two communities, playing off the chorus of an old Broadway song in its title, "The Bronx Is Up and the Battery's Up." After Tim Carey took over the Authority in 1999, however, those connections weakened.

In the New Settlement Apartments themselves, the impression of Battery Park City was less favorable. Local activists' opinions were more in line with those of Adam Green, the founder of Rocking the Boat: that Battery Park City's privilege was an affront to New Settlement Apartments, an injustice considering the ongoing needs of the community, and that the charity directed toward New Settlement in no way rectified the imbalance that remained between the two communities.

Jack Doyle, director of the complex, came to New Settlement Apartments in 1995. Previously, he had been the Red Cross's director of homeless services in New York. Not long after the suburban strategy's homeless sweeps and service cuts increased in 1994 and forced more homeless people out of gentrifying Manhattan and into the Bronx and beyond, Doyle himself came to the Bronx to work at New Settlement.[21] Doyle was shown profound respect by the staff at Settlement Housing Fund and was universally known by activists in the community, who believed he deserved much of the credit for how well New Settlement Apartments was run. He was deeply committed to New Settlement; people who directed me to him gave the distinct impression that he so represented the place that he practically *was* New Settlement (a characterization he would have surely and sharply rejected). Having been told by others how knowledgeable he was about New Settlement Apartments, I had expected he would be interested in my comparison of that complex to Battery Park City. But when I first sat down with him in a meeting room in the New Settlement management offices, his attitude toward me was cool, even suspicious. It turned out Doyle's years serving a low-income community in the Bronx had left him justifiably dubious of Downtown interlopers, particularly those talking about intriguing partnerships between places like Battery Park City and New Settlement. He denied that there was any similarity between the two neighborhoods, or even much of a connection. In a frankly refreshing change from the platitudes of proponents of Battery Park City's financial underwriting of New Settlement, Doyle resented Battery Park City's class privileges. He warmed only when he was convinced his interviewers would not ignore the inequities between the two communities. When Alexandra Demshock, a student with whom I was conducting the interview, mentioned her membership in a local church that had allied itself with New Settlement groups in community struggles, Doyle's attitude changed. Concrete commitments to the community satisfied him; abstract discussions of linkages between rich and poor corners of the city did not. Asked how he saw Battery Park City from his perspective in New Settlement, Doyle responded firmly, "I see it as reflective of a pattern of discrimination that permeates our city." Battery Park City, after all, had "an embarrassment of riches" evident just in the size of its parks budget. He contrasted that abundance with the needs around New Settlement and wondered "how many young people in Battery Park City got stopped and frisked" at a time when the New York Police Department was conducting pat-down searches without probable cause of half a million people in New York per year, disproportionately black and Latino young men in neighborhoods like New Settlement's. Doyle was not critical of people involved in Battery Park City's project to fund affordable housing, but he was wary of any effort to compare residents of the two communities. For him, the residents of New Settlement Apartments did not exist in the reflection of Battery Park City. Though people with more money and power seemed only to consider them as subjects for reforms or charity projects, they had their own independent lives. New Settlement bridled at the patronizing attitude of its more privileged sibling. As Freire observed, charity that can be extended only as long as the inequality it addresses is maintained is hardly charity at all. People involved in the development, planning, programming, and organizing of New Settlement Apartments recognized what Battery Park City's funding had allowed them to do. But like Adam Green, Doyle and others made clear that, because Battery Park City's contributions at every level continued to fall well short of the need at hand, touting their success required disregarding the fact that people were still poor and still faced daily difficulties because they were poor. As Battery Park City sailed on, the teens of New Settlement Apartments would build their own rowboat.

3

Residents, Space, and Exclusivity

THE YEAR AFTER the first residents moved into Battery Park City, the esplanade along the Hudson River opened to the general public. Instead of celebrating the new waterfront park, residents were resigned. "You can't stop the rest of Manhattan from enjoying it," said one man. "The only thing you hope is that they're not going to find out about it too quickly."[1] From the start, residents had developed a definition of their community that was inconsistent with the use of their public spaces by others.

The parks remained relatively isolated from the rest of the city, so residents incorporated that sense of isolation into what community meant in Battery Park City. "I feel the isolation," Maria Crouch, then president of the Battery Park City Owners Association, said a few years later. "The positive side is that you have a real feeling of community because you get to know your neighbors by spending more time in the neighborhood."[2] Here the second stage of the reciprocal relationship between physical design and social organization manifested itself as the physical design of the neighborhood shaped the social meaning of community. The growing parks and other public spaces of the physical neighborhood helped the community define itself in contradistinction to the rest of the city.

Exclusion has been tied up with the meaning of community in Battery Park City since its inception. To understand the interplay of physical design and social organization, I examine the physical elements of the neighborhood, most prominently the barrier of West Street, to understand how physical features foster social attitudes. The interaction of the isolated space, material privilege, and exclusive attitudes led residents to adopt what I call a "spatial definition" of community rather than one defined by shared history, ethnicity, or shared struggles. Space is not an external influence on social reality: it is in constant, reciprocal dialogue with the social, shaped by and shaping people's collective goals. Studying this reciprocation between the physical neighborhood and the social community of Battery Park City demonstrates those interactions.

Citadels are not only the product of globalization, local government, or business elites. As exclusivity is cemented into a community's social and physical reality, it is preserved through the actions of the private citizens who live there as well. Design elements and the space in which social relations occur nurture already existing attitudes and beliefs. Here the built environment reinforced residents' views toward their community and the rest of the city. In this way space became a medium to transmit social attitudes, from developers to residents, from residents to newcomers to the community, and, ultimately, from the community to the world beyond.

The Exclusive Foundation of "Community"

Despite the high esteem justifiably accorded to the term *community* in other contexts (among, for instance, community activists and community organizers), in urban planning it is inherently problematic. Contraposed against "the public," community is a smaller, exclusive group of people. As the architect and critic Alex Krieger has pointed out, "While we use the word 'community' interchangeably with the word 'public,' community involves selection; a distinguishing of those who belong from outsiders. The public, on the other hand, is—or should—encompass everyone."[3] Tendencies toward community can encourage, Krieger argues, tendencies much at odds with the local democracy that the term is meant to conjure up. "The current rush of enthusiasm for the 'community' found in the traditional small town disregards the many anti-public predilections of small-town life." For all its positive rhetorical valence, "community" has a distinctly exclusionary, even undemocratic, aspect.

The feminist theorist Iris Marion Young has argued that making "community" central to the politics of liberation is inherently contradictory. Community becomes "politically problematic" because "those motivated by it will tend to suppress differences among themselves or implicitly to exclude from their political groups persons with whom they do not identify."[4] Young ultimately rejects community as a basis of feminist organizing, embracing instead a cosmopolitan "politics of difference" or frankly "unassimilated otherness" that does not demand or impose unity where it does not exist.

Richard Sennett sounds a comparable warning about the balkanizing effects of the politics of community. Using as an example the strident efforts of the Queens, New York, community of Forest Hills to exclude public housing from their area, Sennett argued that the creation of community happens at the expense of the public, because the process of social withdrawal that accompanies the creation of community both establishes a category of outsiders excluded from the community and measures loyalty to the community in terms of one's hostility toward those outsiders in supposed defense of the community.[5] In such a setting, truly public interaction is lacking, and community interaction takes its stead. To Sennett, when the public sphere is replaced by the misplaced loyalties of community, a "tyranny of intimacy" reigns.

From the perspective of community ethnographers, it is difficult to see community—which, particularly for less affluent people, is a valuable identity along which to organize in opposition to city hall or corporate power—in such singularly negative terms. It may be that community is qualitatively different in rich and poor communities. After all, "exclusion" means something very different in communities that exclude and protect their privileges and communities of the excluded. (Peter Marcuse calls the former citadels, or imperial citadels, and the latter "outcast ghettos" or "immigrant enclaves.")[6] But that distinction notwithstanding, more aggressive exclusion is always immanent in community.[7] That exclusion is all the more socially damaging in a community that, like Battery Park City, has access to considerable resources. Of course, those resources are the added impetus to the impulse to exclude others from the community's bounty.

For these reasons, community, despite its positive connotation in the language of community politics, is inherently tied up with notions of exclusivity. Invoking "community" in Battery Park City should not provide cover for exclusivity but signal the need to investigate that exclusivity. As will be seen from residents' comments, in Battery Park City community and exclusivity were never far apart. And residents enacted both in distinctly spatial ways that provide examples of how space ultimately shapes and is shaped by community. The views of Battery Park City residents displayed the connectedness of community, exclusivity, and space. That connectedness would prove central to conflicts that residents later engaged in.

In the beginning, Battery Park City's exclusivity was both a social and physical reality. Over time, continued construction of the neighborhood left it less physically isolated, but by then the social reality of residents' exclusive attitudes led them to selectively emphasize those elements that were consistent with their social definition of their community. As seen in later chapters, residents would eventually actively harness the physical design of the neighborhood to reinforce social exclusivity. To recognize this process, it is important to examine early exclusive attitudes.

My examination of 111 articles with "Battery Park City" in the title published in major U.S. newspapers since 1980 demonstrates that from the moment the first residents moved into Battery Park City in 1982 they articulated an exclusive quality about Battery Park City. As David Dunlap commented barely a year after the first complex opened in June 1982, "Tenants in the 1,712-apartment Gateway Plaza . . . are feeling rather proprietary these days about the esplanade."[8] Another article reported that "neighbors relish the quiet that outsiders may see as isolation."[9] Yet another described Battery Park City as "an urban suburb."[10] Eleven years later Dunlap found in Battery Park City "an insular quality that attracts those seeking a respite from Manhattan and bothers those who wish it were more public. Whether Battery Park City will ever be fully integrated into the fabric of New York remains to be seen."[11] It is unlikely that these accounts are evidence of journalistic bias against Battery Park City. Just as Dunlap sought to withhold explicit judgment about Battery Park City's "insular quality," journalists who wrote about the neighborhood's isolation often wrote positively and without qualification about other aspects of the neighborhood. My field research corroborated that exclusive views played significant roles in residents' views and activism on local issues. Residents recalled that their sense of Battery Park City as existing apart from the city formed very early in the neighborhood's history. Whatever its charms, Battery Park City has long harbored significant attitudes of exclusion.

Frequently early residents were concerned about changes to the physical design that would make the neighborhood more accessible to the rest of the city. (These changes also made the city more accessible to residents.) Five years after the first residents moved in, a resident hoped that the opening of the Winter Garden Mall, then under construction, would not "destroy the isolation she cherishes." She said "I don't think the change is going to be as drastic as some people fear it's going to be, but I hope that while we get more services, we can still retain some of our privacy."[12]

In his book on Battery Park City, David Gordon recounted a resident association's objections to connecting Battery Park City's esplanade to Battery Park just south of the neighborhood—"exactly the reason," Gordon pointed out, "for planning a continuous waterfront walkway" in the first place. In a community newspaper one resident complained that the new park connection "certainly looks like a written invitation for people to go from Battery Park to the esplanade . . . [though] I'm not saying that we should put a moat around Battery Park City."[13] Opposition to construction and expansion is hardly unique to this community. But Battery Park City demonstrates the way space is reciprocally related to residents' attitudes, for the development of Battery Park City created the isolation that then contributed to residents' preference that further development be equally exclusive socially and physically. New shopping malls (like the Winter Garden) are often welcome in communities that, like this one, have lacked closer shopping options, and so are recreational projects like the waterfront connection to Battery Park. Whether residents welcome such developments or oppose them as they did here is influenced by the physical neighborhood.

Even in more recent years, residents in Battery Park City sensed the same isolation earlier pioneers had described. "It has always felt like the suburbs of New York," said one resident. Another said she had heard her fellow residents call the neighborhood "suburbia in Manhattan." Some even moved here because of this isolation, while others bridled at it.

In a city where it's always hard to find a home, some residents just moved in because there was a vacancy. They too sensed the effect Battery Park City's design had on their community. Christina Molloy sublet a room in 1998. She told me later, after moving away, that she had enjoyed the parks and the esplanade too but had never warmed to the area. "There wasn't really anywhere to meet people," she reflected. "There really is nowhere to go out down there." For her, Gateway Plaza had the atomized feeling of "a hotel . . . For me in any case it was a landing pad. And I didn't really socialize with people in the building at all." Molloy found herself happier in other neighborhoods.

But the neighborhood's very isolation meant that even for residents like Molloy, Battery Park City became a significant part of their life, and they spent considerable time there. She walked to work in the Financial District and ran along the park "all the time," feeling that the water was the best part of Battery Park City. She even went swing dancing every week at the bar atop the nearby World Trade Center. "That was my local bar . . . I was there every Friday at one point." And like almost every resident, she did most of her everyday shopping and caught the subway at the mall beneath the Trade Center. Regardless of their feelings toward the neighborhood's isolation, much of what residents did took place in Battery Park City or just across West Street's pedestrian bridges in the World Trade Center.

Living or working in Battery Park City was undeniably an urban experience. One man whose family relocated from suburban New Jersey into

Battery Park City after September 11 noticed a dramatic change in his life outside work, even though at the time he was busy directing efforts to re-open one of the large offices in the World Financial Center. "I do remember running out to the supermarket [as opposed to driving] and going to Midtown and hopping on the subway all the time. And I thought, 'I'm really living the city life. For the first time.'" He was struck by the lack of private, outdoor space, but his kids adjusted well, he said, feeling that rollerblading and biking down the esplanade was part of an extended summer vacation. He noticed the social effects of so much shared space, as he repeatedly ran into neighbors on the elevator, in the building's gym, and while he was walking, reading, or attending concerts in the park. He recognized being exposed to more diversity, like gay couples in his building, than he had experienced in the suburbs. The effect of space on the social experience of Battery Park City was not only to create an exclusive community but, compared to truly suburban spaces, to create conditions for more social interaction.

Design Influences on Community

Three physical elements of Battery Park City's design in particular contributed to residents' attitudes toward the community and neighborhood. The defining feature in this regard was West Street, which, along with the Hudson River on the other three sides of the neighborhood, isolated Battery Park City. Inside the citadel, the discontinuous street grid and the protracted construction period in Battery Park City have contributed, respectively, to the sense of isolation and the underlying concern that that isolation could soon be lost. All three factors demonstrate how physical features and the overall spatial context influence the characteristics residents ascribe to their community. While these factors do not demonstrate intentional design exclusion on the part of Battery Park City's planners, they show how a community's shared spatial experiences can contribute to the "collective memory" that members come to share.[14]

West Street

West Street is the most significant barrier to Battery Park City. During most of the time Battery Park City was being planned, it was an elevated highway like FDR Drive directly across Lower Manhattan along the East River. (The collapse of a section of the elevated roadway in 1973 motivated the city to relocate the West Side Highway to ground level.) Today West

Street forms the legal and conceptual eastern boundary of Battery Park City. While it is a surface street with occasional stoplights, it has a highway's scale and effect on the city fabric and is considered by New Yorkers to be part of the West Side Highway that it joins.

West Street is a formidable barrier to pedestrians. Not only do its ten lanes, divided by islands and high concrete barriers, make it much wider than almost any street in Manhattan, but it lacks lights and crosswalks at many intersections, reducing the number of points of access and forcing pedestrians to go hunting for a way to cross the divide. While a mile-long avenue in Manhattan's grid would have twenty intersections and forty crosswalks, this mile of West Street has only six cross streets and ten irregularly spaced crosswalks. Pedestrians approaching Battery Park City need to plan their approach several blocks in advance in order to reach West Street at a point where they can cross. Without prior knowledge of cross streets, it is challenging to find a way into Battery Park City.

Before September 11, the easiest and even most common way into Battery Park City was not at street level at all but via the pedestrian bridges that went directly from the office buildings of the World Trade Center complex to the World Financial Center. But the pedestrian walkways appeared to be private: they were generally accessed through office buildings and were quite easy to reach from those buildings' lobbies but almost impossible to enter from West Street itself. Crossing them during public space research I conducted at the World Trade Center and Battery Park City before September 11, I noted my own sense of unease entering the doors of an office building and going down a wide, quiet, enclosed walkway in order to cross the street. Below the bridges, at street level, finding a way to reach Battery Park City or enter the Winter Garden Mall was even more difficult than it is today. The bridges made an easy trip for businesspeople but hardly one that seemed intended for the general public, so they did not overcome the problem of public access to Battery Park City.

Residents commonly described West Street as a barrier and discussed its effects on their community. The first effect was to create a place apart. As one woman observed, it was "an escape. It's quiet. We don't have cars within Battery Park City, really. It's just an amazing feeling. You just get off West Street, and you're in another world." Mark Watkins, who had recently had to move out of Battery Park City, had a similar description: "The West Side Highway acts sort of like a wall, and it's just really quiet here. Lots of families and babies and dogs. It's pretty." Perhaps an outsider would be perplexed by residents praising a neighboring highway as a source of quiet.

In this photo (ca. 1988), the gap between the World Trade Center (*left*) and Battery Park City (*right*) is bridged by a pedestrian walkway over the many lanes of West Street. That highway's few, widely separated crosswalks and brief "walk" signals at traffic lights contribute to Battery Park City's physical isolation. (Note the Cadillacs, Lincolns, Rolls Royces, and cluster of taxis waiting to pick up passengers in the driveway of 1 World Trade Center.) (Photo: Robert Simko)

(In fact the highway is quite loud; it keeps out not cars but pedestrians.) This oddity speaks to the relationship of physical features to social meaning. Any number of aspects could define the meaning of a physical feature like West Street. Had the community had a strong sense of being well connected, residents might have mentioned how the fast-moving street connected them quickly to the rest of Manhattan. Had the community felt the city neglected their well-being, residents could have pointed to the dangerous street outside their door. Had they felt stuck in a crowded commercial corner of Manhattan ill suited for a residential community, they could have pointed to the noise and truck traffic. No one meaning adheres to the street, so residents' feelings about their neighborhood interacted simultaneously with the physical conditions around them to establish a social description of their community that they could see reflected in the physical features of the neighborhood. The fact that in countless discussions with me about the design of West Street no one complained about the highway next door speaks to the overriding value of its "wall" function, and the depth of this shared conception of Battery Park City's relation to West Street.

In contrast to residents, designers readily recognize that Battery Park City retains a tangible isolation. "It's an enclave," said David West, an architect whose firm designed two buildings in the North neighborhood of Battery Park City. "The project essentially turns its back to the city," conceded Stanton Eckstut, one of the main architects of the 1979 master plan, though he argued that this was because at the time of the design process the street, then called the West Side Highway, was elevated.[15] Sandy Frucher, former president and chief executive officer of BPCA, concurred: "West Street is a horrific physical barrier, to Battery Park City in particular, and it creates a sense of isolation that is in some ways unforgiving." West Street not only bounded the community but set it apart.

Residents recognized that the barrier of West Street kept people, cars, and the rest of the city at arm's length. "Even though it's five minutes away by walking, I do feel like there's a barrier of some kind that the West Side Highway creates," said Mark Watkins. We were talking in the modernist lobby of his building, which had views of the water through its high glass windows. When I asked what he thought about people who saw the barrier as a good thing, he answered, "I can see it both ways. The barrier, that's what makes the area feel separate and safer. There's no traffic. Very little traffic. The people who are here are the people who live here. It's not rowdy, and part of it is because that barrier is there. The idea to maintain that is kind of silly. If you live in Manhattan you should be connected to it." Like other residents, Yarrott Benz denied it was a barrier but provided examples of how it actually was, along with other elements that contributed to the sense of separation: "Well, I don't think it's West Street that's the barrier, it's the fact that after you cross West Street it's another half block before you get to a street that has the lobbies on it. On the other side of West Street, it's all commercial. So there's this big buffer between West Street and your house. If all those apartment buildings that back up onto West Street and open on South End Avenue, if they opened on West Street—I'm not saying they should—it would feel a little closer. I'm not saying it should change, because I kind of like the separation. It's kind of nice to have both flavors." Like Benz, others made it clear that they liked the sense of separation West Street provided their community. Sitting at a table on a second-floor balcony of the sunlit Winter Garden Mall atrium, Community Board member Anthony Notaro, looking down upon the community in motion all around him, commented on residents' views regarding the neighborhood's isolation:

> The word *isolation* is not as appropriate as *boundary* . . . 'cause *isolated* says you want to come, you feel separated and you want to join. I'm not sure, it may be the opposite. Sometimes I feel it's a boundary: "I don't want the rest of the city to encroach here." I don't necessarily agree with that, but sometimes I get that feeling. You hear a lot of people say they really didn't want the access improved between Battery Park City and Lower Manhattan. And there are times when people really want to keep it separated. I don't know that that's prevalent. You know, it's not like walking down Third Avenue, and you can go down three different neighborhoods. You're here, and then you have to go to Tribeca. And that physical characteristic sets people to have a certain perspective here.

Residents sensed that their social community was in this way influenced by the physical design of the neighborhood and that it enclosed them in a way that many wished to preserve. As Notaro suggested, it was less accurate to describe Battery Park City as isolated (meaning "kept away from others") than as exclusive ("keeping away others").

In addition to the shared interpretation of the highway as a source of solitude, West Street had several effects on residents' definition of community. First, it provided a common delineation of the boundaries of the community and the neighborhood, since all the land west of West Street was administered by the BPCA. Thus in Battery Park City, unlike other communities, there was near-universal agreement about the boundaries of the community and the neighborhood, and the two coincided. The congruity between the community's distinction from other places and the physical sense that residents were cut off from the rest of the city reinforced residents' sense that community was the appropriate scale at which to define their residential lives, and that community should be defined spatially. These residents became invested in keeping the barrier and would later seek to defend it against planned alterations. Residents did not fear what was on the other side but were unhappy about who might get across the barrier. No one discussed this potential breach with me in terms of the most common American fears of outsiders—crime, racial difference, threats to class status, or danger.[16] Instead, discussions were directed at "tourists," "crowds" in the park, and nonresidents using the public amenities. Tropes of racial difference, crime, poverty, and the "danger" of the lower classes have been excavated in studies of working-class and middle-class communities (particularly white ones) that seek to distinguish themselves from poor communities and/or communities of color. But upper-middle- and upper-class Battery Park City was neither socially proximate nor physically proximate to low-income neighborhoods, black or Latino communities, or high-crime areas. Thus they felt less urgent social pressure to distinguish themselves from the most disenfranchised, who were socially distant from themselves, and more need to distinguish themselves from the middle class and the economic peers with whom they competed for status: tourists and other relatively affluent people from Downtown who used the park.

Most critics have considered the barrier of West Street both a symbolic and an intentional means of separating Battery Park City from the rest of the city. The assumption of concordance between the intended and actual function of West Street squares with the tendency of postmodern writers to read the city as a "text." The history of West Street's development as a privatizing barrier demonstrates that even spaces that seem so unambiguously symbolic—here, of exclusion—have more complicated histories. The particular history by which West Street came to serve as a socially significant barrier for residents provides a richer understanding of the contingent manner in which physical designs are employed for social ends.

To assess how intentional the barrier function of West Street might be, it is worth considering the justification in the 1979 master plan for one of West Street's irregularities—the private pedestrian bridges that carried most people in and out of Battery Park City. The planners wrote that unlike earlier planners they generally didn't want to use pedestrian bridges in Battery Park City, so traditional streets and sidewalks predominate. But at West Street, the broad and busy highway merited an exception, in the form of elevated bridges to reduce the danger to the large numbers of pedestrians who would be crossing it. "This level of pedestrian movements [forty-three thousand per hour] justifies the provision of a major elevated pedestrian link in order to cross West St. and Westway. . . . Elsewhere in Battery Park City the projected pedestrian flows will be less and the provision of elevated pedestrian circulation systems are not warranted except where crossing of Westway [West Street] cannot be easily accomplished at-grade."[17] The significance of these pedestrian bridges should not be underestimated, since their effect was to filter access to Battery Park City. Steven Flusty describes spaces like Battery Park City as "slippery," in that they "cannot be reached, due to contorted, protracted or missing paths of approach," and notes that they often deflect criticism because their interdictory effects can be blamed on preexisting conditions (like West Street).[18] But while skeptics could call the safety rationale mere cover for a desire to privatize the entrance to Battery Park City, the pedestrian risks crossing West Street are real. Planners sought to mitigate, not exacerbate, the barrier function of West Street.

More could have been done to diminish the barrier effect of West Street in subsequent years. Though there have been long-term renovations of West Street, it has not been reconfigured, from pedestrians' perspectives, as a more conventional city street. Part of the problem is that, as a highway, West Street is maintained by the state, not the city, and state plans for it have clearly reflected the auto-centric perspective that has shaped the rest of the state's highways. Battery Park City buildings have done nothing to help: virtually no Battery Park City buildings come to the edge of the West Street property line. Rather than facing the street like other New York office buildings, the World Financial Center is buffered from West Street by a slope of grass. In fact, along the length of West Street such spaces, along with vacant lots, have amplified West Street's impact as a no-man's-land and dividing line.

The effect is to make West Street seem not like all the other city streets one crosses, but like a suburban access road that only cars, not people, should use. Yet although the BPCA took little initiative in twenty years to improve public access across West Street, there is still conflicting evidence regarding whether the Authority wanted West Street to function as a barrier: the Authority and the major commercial landlord both appeared to support post–September 11 plans to bury West Street, covering it with a smaller, greener street that they believed would be easier to cross. As discussed later, the major justification for this plan was that it would have better connected Battery Park City to the rest of Manhattan. (Residents, on the other hand, opposed the tunnel and wanted to design West Street in a way that maintained its barrier role.) Thus the evidence that influential actors wanted West Street to isolate Battery Park City is largely circumstantial, while important counterevidence, like post–September 11 efforts to submerge West Street and planners' prioritization of safety rather than exclusivity, demonstrates at the least that there was no singular objective regarding West Street among elite decision makers. West Street became a physical obstruction through a combination of incidental developments (like the collapse of Westway), the failure of planners to affirmatively redesign the space (as architects and planners set World Financial Center buildings further back from the street), and widespread acceptance by planners and Authority officers of Battery Park City's physical isolation as part of the

Battery Park City lies on the mile-long strip of land west of West Street. The barrier role of West Street can be in part explained by a glance at the surrounding area. Not only were the six to ten lanes of West Street a barrier in themselves, but for many years an unusual quantity of block-size vacant lots abutted the street down the length of Battery Park City. Additionally, the lighter gray shading indicates buffer areas, including grassy lawns, bike paths, and Little West Street at the south end, which was fenced and closed. Key: Dark shaded area: vacant lots, no access. Lighter shading: park areas—these are pedestrian accessible, but they make the West Street corridor broader.

facts on the ground. In this manner, a physical feature that people recognize as an unambiguous symbol of social reality—as West Street symbolizes Battery Park City's isolation—can, as it does in this case, have a genesis that is far more contingent. But once West Street adopted, however unintentionally, the role of a social barrier, it was not the Authority or global capitalism's architects that ensured its continued service in that role, but residents themselves.

The Grid

One of the few things Battery Park City's planners and critics do agree upon is that the neighborhood reproduces the accessible street grid of Manhattan. In an early, influential critique of Battery Park City, Julia Trilling wrote that Battery Park City "proudly keeps the street grid, so representative of New York. But it does so in an area of New York—Lower Manhattan—that never before had a grid."[19] Often critics dismissed the grid as dishonest historicist appropriation. But observers thought there was much more of a New York City grid to Battery Park City than actually appears on a map. Master planners did not seek to recreate a conventional street grid. Battery Park City's streets have distinct effects on the social form of the community, but we must consider both their actual form and planners' rationale for their design.

Battery Park City's street plan varied from the Manhattan grid in several ways. The streets were named, not numbered, included nine cul-de-sacs, and formed irregularly shaped blocks. Most streets did not physically cross West Street or allow for pedestrians to cross it There was no north-south avenue running the full length of the neighborhood, so it was actually impossible to drive through from one end of Battery Park City to the other. Despite sidewalks, crosswalks, and familiar streetscapes, the experience of navigating Battery Park City was quite different from walking the gridded city and was widely described by newcomers as disorienting.

The effects on the community of the modified street grid are more subtle than those of West Street. For instance, the lack of north-south streets, in particular between the North and South neighborhoods, made its mark on residents' networks. Residents in the South neighborhood said that although they knew people in other buildings they had few contacts in the North neighborhood. Mark Watkins observed that in addition to West Street, "There's another barrier they can't fix. South End and North End. They dead end at the World Financial Center. And that creates two Battery Parks. And that's a shame. Because the South End and North End would be a lot livelier, and the people would be maybe more engaged with it, if it was a straight shot. And maybe some kind of connection over the West Side Highway might alleviate that." Debra Lee Murrow of the northern neighborhood concurred that North End and South End were like two different communities, and local activist John Dellaportas said that when outsiders talked about connectivity in relationship to West Street, they should have instead considered connectivity between the North and the South neighborhoods. To the degree that Manhattanites had difficulty understanding the layout (in contrast to the instantly readable grid of the middle of the island), they were less likely to use it or find their way around. But whether or not there were more through streets, Battery Park City itself did not lead to anywhere else (since it projected out toward the water), making access for most through-traffic moot.

However, this lack of a grid did not spring from designers' intention to make Battery Park City unwelcoming. Considering that every previous plan had intended to completely break up the streets, elevating pedestrians and hiding cars below the surface, the 1979 master plan was a step toward easy readability of the space and compatibility of the design with street plans that New Yorkers would recognize and easily understand. More importantly, the Authority's design guidelines made it clear that they did intend Battery Park City's streets to be less accessible, but they explained why: "Auto access is provided to all of the blocks; but the discontinuous street pattern discourages through traffic and fast speeds."[20] Meanwhile, "pedestrians can walk along the Hudson River edge without having to cross the paths of automobiles."[21] Contrary to what might be expected from a market-oriented project, people were funneled onto a noncommercial promenade and away from street-front retail. The grid represented the low-level privatization found in the plan rather than intentional opposition to public use. The space was designed with other objectives in mind and lacked a strong commitment to accommodating the nonresidential public. Such an approach may be guilty of neglecting the public sphere, but it is far from the attitudes that inform typical gated communities.[22] In more ways than one the neighborhood did not embrace the city, but it did not shut out the city as completely as it could have.

Piecemeal Construction

Piecemeal construction accounts for the empty blocks Yarrott Benz described as bordering West Street and contributing to its quality as a no-man's-land. While the effect of thirty years of ongoing construction is less striking now that the majority of Battery Park City has been built, the effect was more pronounced until recently. During that time, the core of the community was set apart from the city by empty land and was regularly altered by temporary access routes. Vacant lots widened the effect of West Street because those inland lots were the last to be developed. Waterfront lots were developed first. Construction continued to block access and travel down neighborhood streets. Like the alteration of the street grid, this piecemeal approach, while used for completely unrelated reasons, further separated Battery Park City from the rest of Lower Manhattan.

Socially, the effect of prolonged construction has been to bestow on residents the universally shared experience of moving in and becoming accustomed to a relatively secluded area and then watching as that area becomes less and less so over time. A narrative of lost seclusion is common to residents' accounts of the community and informs how they see others' use of the public spaces. (Ironically, of course, the construction suggests that most of the new users in the parks have been fellow residents. But since residents and visitors are likely to be indistinguishable, older residents always assume the new crowds are outsiders.) Because every resident to arrive in the past twenty-five years has experienced this loss of seclusion, it has become a part of what sociologists would describe as Battery Park City's collective memory. *Collective memory* usually refers to the way citizens conceptualize and commemorate their nation. But the same process can occur at the community level as well. Collective memory does not simply reinforce existing lines of inclusion and exclusion. Instead, "Part of the struggle over the past is not to achieve already constituted interests but to constitute those interests in the first place."[23] The process of describing the loss of seclusion, like the process of defining West Street as a boundary, has been central to identifying what is included, what is excluded, and what has been lost.

After growing accustomed to the neighborhood, residents have been reluctant at each point when they're expected to give up a little more of their seclusion. Linda Belfer, one of Gateway Plaza's first tenants in 1982, reflected on how piecemeal development made residents feel that something was being taken away from them. "I was here maybe three years before they started building other buildings. And we were very resentful: How dare they start building other buildings? And we knew from the get-go it was going to be a whole development. But your brain doesn't recognize that, and we got very angry."

Sandy Frucher, former head of the BPCA, told a similar story, in less sympathetic terms, about a resident who came down one morning in her bathrobe and sat in front of a bulldozer, demanding that it stop. She cherished the view of the Statue of Liberty from her window and was outraged by the injustice that it would be taken away from her. Frucher used this anecdote as evidence that residents of Battery Park City—like those of any subsidized development project, he claimed—had an inflated sense of entitlement. His perspective is probably colored by the fact that he would have been the target of much resident dissatisfaction while he was at the Authority, but the ongoing experience of having views, sunlight, quiet, and space taken away as more buildings went up (even as more amenities were provided) nonetheless influenced the story Battery Park City residents told about their community. Conversely, piecemeal construction has also meant that critics who wrote about Battery Park City in earlier stages of its development found it more exclusive than it is now: Phillip Lopate described guards in booths at the southern entrance of the park who barred nonresidents from entering after dark.[24] The park is public, however, and the booths and guards are now gone. Battery Park City's slow development thus simultaneously heightened residents' sense of how isolated it should be and outsiders' sense of how isolated it was, firmly establishing public perception.

In these ways, even though Battery Park City was not built to be the impenetrable fortress it is portrayed to be, design elements contributed to residents' sense of isolation and exclusive conception of their neighborhood. Of course, exclusivity is found in other communities as well, and space probably plays a role in those cases, too. The built environment, by facilitating or inhibiting access by other people, influences the social construction of relationships, groups, and communities. As the influence of programmatic exclusion has already shown, space alone does not define a community, and the successive decisions about and by the people who will be there play a large role. But through the reciprocal relationship space is a peculiar catalyst, established under a certain set of social relations and intended to help reproduce those social relations indefinitely. Battery Park City's carefully created spaces make the role of space more pronounced, and residents reflect that influence in their description of the community.

While social processes lead to the shape of a particular space, the shape of a particular space cannot, by itself, reveal in its physical form the social processes that gave rise to it. Space, history, and social relations must be examined together. Social processes and history interact to form space, so that the historical variables disrupt a direct relationship between space and social relations. As Manuel Castells observed, "Space is not, contrary to what others may say, a reflection of society but one of society's fundamental material dimensions."[25] The history of a space's creation is necessary for an adequate understanding of its production and intended meaning. Understanding the citadel, the key to the global city, requires careful examination of the history, space, and social relations found in the project. Doing so provides considerable insight, not only into the citadel, but into the type of city to which it belongs.

The Spatial Definition of Community

Not only did space reciprocally reinforce social conceptions of the community as exclusive, but the spatial and the social acted in tandem to produce a distinct definition of community. First, residents described their community in markedly spatial terms and celebrated the neighborhood's public spaces before anything else. Second, in contrast to other neighborhoods, residents did not use shared history to define membership in the community; in Battery Park City, community was defined by a spatial model.

Residents consistently lauded Battery Park City for its public spaces. To explain why they had moved there or had stayed, they mentioned the parks, the esplanade, and the baseball diamonds (even when they didn't play ball). This wasn't inevitable because Battery Park City had many other attributes to recommend it: the livability of the neighborhood, its convenience to downtown and the rest of Manhattan, the friendliness of residents, the age mix. Pearl Scher, who could also have focused her praise for Battery Park City on how it had allowed her to remain politically active beyond her retirement as a Westchester County legislator, on her many friends in the neighborhood, or on the neighborhood's cleanliness (she took it upon herself to get business owners on the other side of West Street to clean their sidewalks), explained what was distinctive about Battery Park City this way:

> It's like the finest European resort areas. The esplanade is matchless, and I don't care, I've been to Europe. The esplanade is gorgeous. It is an unbelievable place. Unbelievable because it has the river on one side, and all that park area on the other side. So that when you walk down the esplanade, you're in a heavenly place, it's as if the world doesn't exist outside the esplanade. I love it.

But there are some other things. There are the ballparks, there are the little things. I'm sure that when the teardrop park is finished, it is going to be incomparable. It is a special park, because it is an attempt to recreate what you might have seen in Westchester County or Ulster County, which is a suburban or rural side.

When asked what was different about Battery Park City compared to the Lower East Side, Mark Watkins also began by talking about the solitude. Like Scher and many others, he compared it to the suburbs. In Battery Park City, "it was much nicer, and peaceful and quiet. It felt like the suburbs. That [the Lower East Side] felt like the city. It was very loud, there was garbage and trucks, and alarms, and a lot of noise. [In Battery Park City] it's like not even being in Manhattan." Community Board member Anthony Notaro saw Battery Park City's characteristics in "not just the architecture, [but] the open spaces, the maintenance of the parks, the children's programs that the Parks Conservancy puts on." When asked what Battery Park City was like, Yarrott Benz talked about the quiet. "It's very pleasant. Friends of mine who visit say it's like visiting St. Paul [Minnesota]. Its so quiet and when I walk my dog, it's dead, it's so quiet. Sometimes I feel like it's Venice, because when I have my windows open, the noise you hear is the noise off the water." Benz, like Scher, alluded to European public spaces. One of the reasons he moved to Battery Park City was "the views and the sense of space. The plazas—the Marina or North Cove—it's the closest I've come to the Piazza San Marco." For both of these residents, the community was defined by its public spaces.

Residents felt that the public space was their own private space. Abby Ehrlich, who planned events for the Battery Park City Parks Conservancy, recounted how "for two or three years I've had a woman come down to story time [in the park] and say, 'I don't want this in my yard.' I had a man come down during dancing and say, 'I don't want this in my living room.' This one guy said—and there were twelve people there—he was screaming at me. He was cussing me out. He didn't like the music. He told me what kind of music to have. I told him thank you, we can't afford the Rolling Stones." Because the parks defined an inwardly turned community, they could shift from being conceived of as space for the public to being more restricted, community space. The centrality of space in these definitions is in contrast to many of the other things used to define other communities.

One irony is that while the parks are of consistently high quality in Battery Park City, residents are mistaken in their impression that there are a lot

Battery Park City's esplanade, with its nineteenth-century-styled lampposts lining the waterfront walkway, reminded residents of promenades in Europe.

of them: while there are 282 New Yorkers for every acre of park citywide, there are 250 Battery Park City residents for every acre of its parks. This is not much more space, and by the time Battery Park City is fully built, it will actually have less park space per resident than the norm citywide.[26]

That residents prioritized public space when discussing what they liked about Battery Park City reveals two different aspects of the community. First, residents in this high-rise urban setting used public space much more than their upper-income peers who lived outside big cities. Battery Park City is different from most affluent communities in the United States. Most often, affluence means that people have the means to acquire privately things that are public goods to many of us: private schools instead of public ones, private clubs instead of public parks, private cars instead of public transit, private backyards instead of the street corner. Despite the wealth of most of Battery Park City, high-rise living there meant that the area's public spaces still played a central and meaningful role for residents for recreation, socializing, and even community organizing. Much of what I observed Battery Park City residents doing in public spaces their peers elsewhere do in private, from sunbathing to playing with the kids to eating dinner (whether at the outdoor seating of a local restaurant or on a picnic blanket in Rock-efeller Park in front of a summer performance).[27] Thus while Battery Park City's community defined itself more exclusively than other New York communities, it still used public space far more than most upper-class residents did outside the city.

The second rationale for the greater importance accorded to space (and the lesser significance accorded to strictly social aspects of community) is that compared to communities that aren't wealthy, Battery Park City residents were less reliant on the social contributions a community could make to their well-being. For instance, residents here had less need for neighbors to perform child care (either regularly or in a pinch) because they could pay for other caretakers. Upper-class friendships are rarely as locally concentrated as those of working-class people. And residents certainly spent less time in the neighborhood than adults featured in studies of working-class "street corner societies" by William F. Whyte or Elijah Anderson.[28] Battery Park City residents described their community in terms of its public spaces because community spaces were *more* important to these residents than to people in other upper-income areas, and because the interpersonal contributions of community were *less* important to affluent residents who could buy those services instead.

But residents didn't just *describe* their community in terms of space, they *defined* it through space. People who live in other places often use something other than space to define their community. In his study of an African American working-class New York neighborhood, Corona and East Elmhurst in Queens, Steven Gregory described the central importance of a shared history to the definition of the community: "It was at events like the Civic Association's dinner dance that the concepts of community, identity, and culture seemed most concrete to me, most dynamic, and most insinuated in the practices of politics. For it was by recollecting past struggles and achievements, producing shared meanings about the present, and dancing the electric slide that residents of Corona and East Elmhurst constructed a black community with complex and deeply historical commitments to social activism."[29] The description of community as most tangible when enacted through shared events resonates with my observations at Battery Park City's winter tree-lighting ceremonies, group meetings, community performances, and memorials. But Gregory focuses on instances, such as a community dance and award ceremony, in which residents used history, more than anything else, to construct, preserve, and define the community: "The Civic Association's annual dinner dance was an occasion when

Corona-East Elmhurst made itself a community by recollecting people, events, and meanings from its past and fashioning them into a shared, yet far from seamless, vision of the future. . . . When John Bell received his Pioneer Award for community activism, he recounted the contributions that had been made by John Booker, Jake Govan, and others who, up to their deaths, had stressed the importance of being vigilant and informed about struggles for social justice."[30] The ceremony also included awards to three people who had grown up there and had since moved away but were still considered members of the community. Not only did the past help define the community, but past residents remained part of the community. Gregory describes people placing someone they met into context by establishing their past and their history of personal connections in the community. Similarly, Sudhir Venkatesh describes how in public housing towers (that were not much older than Battery Park City) "in many invocations of 'community,' tenants repeatedly made reference to the past."[31] Residents in the Canarsie neighborhood of Brooklyn selectively used tradition to counter social changes to which they objected and which they saw as threats to the community.[32] In every case, residents used history to construct and maintain their community.

In the New Settlement Apartments, the Bronx neighborhood built with surplus Battery Park City funds, community was defined by social connections. In my conversations there, the community's most active members, at least, defined membership through participation in shared struggles: other parents who had also mobilized against systematic inadequacies at the local elementary school were part of the community, whether they lived in New Settlement buildings or not. Rocking the Boat retained strong ties to the community even after it was no longer housed there. The importance of shared history was compounded by the long tenure of most residents.

In contrast, Battery Park City residents almost never referred to history. Several residents told me about how many people had moved out of their building because of September 11, but although half of Battery Park City had moved away in the two and a half years since September 11, 2001, no former residents were mentioned by name in interviews or meetings I attended. Activists spoke of how many vacancies there were among the officers of community organizations, and how many times the leadership had turned over because of the rush of moves, but they never mentioned who those people were or what they had done as activists. September 11 is, of course, the common exception to the rule that history is not mentioned in Battery Park City, since it is frequently invoked by residents. But even in this context, the aversion to calling up the names of past members applies. On both the first and second anniversaries of September 11, the *Broadsheet,* the community paper, referred in its memorial coverage to eight Battery Park City residents who had died, but in neither case did it name them.[33] In contrast, the paper tries to run birth announcements, with photos, of every baby born in the neighborhood.

In Battery Park City, construction of a shared political history of the kind Gregory and Venkatesh found was dismissed. On occasions when Linda Belfer tried to motivate fellow members of the Battery Park City subcommittee of the Community Board with claims to Battery Park City's earlier commitments to affordable housing or its past tenant struggles, her comments were met with silence rather than recognition. In a meeting of a larger group of Community Board members (including those from other Lower Manhattan neighborhoods) about plans for West Street, several members who lived elsewhere invoked the memory of the heated "Westway" battle over the same highway some fifteen years earlier. Community Board member and Battery Park City resident Anthony Notaro signaled that the past was unimportant when he responded, "I don't know *what* Westway is. And you know what? I don't want to know." He was not hostile to the speakers' position, but he did not want past struggles to become a touchstone in the discussion. In Battery Park City, contrary to other places, the people most active in the construction of community rejected the relevance of community history to that process.

While people in other communities were able to identify each other as members of the same community by naming shared acquaintances, history, and experiences, in Battery Park City residence alone legitimized one as a member of the community. Of course, other factors could lead two people to feel they had more in common: perhaps both had lived in the neighborhood before September 11 or thought they had seen each other around at the playground. But none of this was a prerequisite for community membership.[34]

Instead of shared history, Battery Park City residents used what I came to call a spatial definition of community membership. Attributes of the spatial definition of community—that membership is based on "being there," lasts as long as one stays, and terminates upon departure—were reinforced through their similarity to membership in other institutions common to Battery Park City residents' lives. The spatial model most reflects membership criteria of the white-collar professional's workplace: in a large corporate office, people arrive and depart, and are bona fide members as long as they are there. As bars around the World Financial Center show, groups of co-workers may socialize with each other, even frequently, but rarely do those interactions continue after someone has stopped working there. A company presents an award to a current employee, never to one that has left. (Like Corona, Battery Park City presented an award at its annual block party, but it was to Linda Belfer, a current resident. No former residents were so honored.) Employees rarely find meaning for their workday in what a company did twenty years ago. In these ways, the spatial criteria for community membership were consistent with the criteria for membership as a corporate employee, and the analogy is useful for understanding the concept of community that residents used.

Similarities between the structure of the community and the white-collar workplace highlight class differences in the meaning and demands of community membership for upper-class residents in general and Battery Park City residents in particular. Residential and job mobility are easier for people with more resources, so one's community can change frequently and with relative ease. Studies have shown that poor and working-class people rely on networks of kin (real and fictive kin) to get through life.[35] Most people rely much more than wealthy residents on the proximity of family and close friends for child care, financial assistance, sociability, and security. While some uncertainty can always accompany changing jobs, the process is similarly less dramatic for typical Battery Park City residents than for workers for whom seniority is the main source of security or raises, or for whom general financial insecurity is greater. Thus the models of community and work show important parallels: new members join the group merely by being present (though the class and ethnic segregation of both environments plays a role), members can just as easily exit, and both entrances and exits happen frequently.

The spatial model of community correlated neatly with the delineations of Battery Park City's particular privileges. The largesse of the neighborhood's PILOT tax system and the resources provided by the BPCA both accrued to residents on that side of West Street. Residents enjoyed these benefits not as a result of their history with the community but simply by living within its boundaries. The exclusivity that residents felt so acutely was also aligned with a spatial delineation of community. The particular privileges associated with life in Battery Park City encouraged such a sense of exclusivity among residents (and that exclusivity became intertwined with the spatial forms of the neighborhood, such as West Street). The combination of distinct neighborhood privileges with social and spatial exclusivity led residents to develop a spatial definition of community rather than a definition based on other factors like shared history. In these ways, spatial and social realities became inextricably linked for residents, and, as subsequent chapters show, residents mobilized to protect spatial features that reified privileged social relations.

The spatial model of community used in Battery Park City is important because of its implications for determining who was included and excluded from the community, and is interesting because it alluded to, but was distinct from, other means of delineating community. The definition involved more than mere propinquity, since some households on the other side of West Street were just as close but were not part of the community. The spatial definition was facilitated by the relative socioeconomic homogeneity of the community but differed from it because residents downplayed class differences among those who were "in," at least in public settings. Thus the spatial definition of community partially coincided with other potential ways of identifying community but for residents set in motion different operative beliefs about the Battery Park City community.

The Reciprocating City

Broad exclusion, both by elite planners and by upper-income residents, is a key feature of the vision of the contemporary city that Battery Park City provides. In the case of Battery Park City, the reciprocal relationship ends up exaggerating tendencies of the upper class. The history of Battery Park City shows that non-elites were unable to effect permanent changes to the programmatic plan for Battery Park City that would have made it more inclusive; thus it represents a failure to implement broadly inclusive planning. Instead of a democratic process in which the Hudson River landfill was disposed of in ways that addressed the needs of a wide range of user groups, the process was one of elites planning the community for elites. Because of the reciprocal relationship between people and space, this meant that elites developed a spatial and programmatic plan that articulated their own priorities. When residents of similar socioeconomic status moved in, the design reinforced tendencies toward exclusivity and segregation and seemed to rationalize material advantages enjoyed by Battery Park City residents. Like feedback from the speakers of a punk rock concert, elite control of the reciprocal relationship amplified the most shrill notes in the rich chords that could otherwise define an urban upper-class community.

The broad economic and de facto racial exclusion, and the distorting effects of an elite-controlled reciprocal relationship, have relevance well beyond Battery Park City. The assumptions that lead to both are widespread: that land use should be assigned in large, economically homogeneous swaths, and that land use decisions can be made by elites (large developers, corporations, and supportive politicians) without significant input by residents of the city at large. While all of New York will never look like Battery Park City, current projects elsewhere around the city are consistent with Battery Park City in their economic and racial segregation and the consistency with which they meet elite, but not public, needs.

The fact that the community is defined spatially could mitigate such reciprocal feedback if it included within those spatial boundaries a more diverse group of residents. With the financial success of the 1979 master plan, the BPCA had the means to provide such diversity. That they did not provide it was a lost opportunity. The fact that the space is segregated economically and is not integrated racially means that the boundaries are not just spatial but, in reality, economic and racial as well.

Spatial definitions of community do not need to have such an effect. To the contrary, the cohesion produced by a spatial definition of community could draw together members of a community that had been programmatically planned to have a more diverse economic or racial profile, particularly if the state provided residents of such an integrated neighborhood with resources as generously as it did Battery Park City. In such a case, the potential differences based on income or ethnicity would be to some degree counteracted by the shared sense of membership within a spatially defined enclave.

In the absence of such integration, the class exclusion of Battery Park City provides more insight into the nature of the suburban strategy taking hold in New York. It is one in which even the middle class is dramatically marginalized, and in which the upper middle and upper classes of places like Battery Park City are the people for whom the city is explicitly organized, at great expense.

Space, privilege, and exclusion thus reciprocally informed the community but did not determine what it would become. The spotlight that shone on Battery Park City after the destruction of World Trade Center starkly illuminated both the exclusive social and physical structure of the citadel and residents' own efforts to reshape their community. Space continued to be centrally important, but residents took an active role in shaping the physical neighborhoods as well as the social community.

4

Oasis to Epicenter

Battery Park City in the Year after September 11

MANY RESIDENTS WERE in their apartments the Tuesday morning that planes struck the towers of the World Trade Center. Like thousands of Battery Park City residents, Eleanor Rosen, who had moved to Battery Park City with her late husband when they first came to the city twenty years ago, could see the towers from her window.[1] But she first realized what had happened when her niece called around nine. "I was restless. It seemed awfully noisy, noisier than usual. But how would it occur to me that this would happen?" she asked. "Soon after that one of my neighbors came by and said, 'What should we do?'" Then the towers fell, and everything went dark. "You couldn't see a foot in front of you [from clouds of dust.] Every leaf of grass, every leaf of the tree was covered with this gray—I don't know what you call it—pulverized building material. We went down to the lobby, and they advised us to walk south on the esplanade . . . What struck me when walking south on the esplanade, what really hit me were shoes. *Shoes* scattered on the esplanade. I don't know why it made an indelible impression on me. And paper."

Along the waterfront esplanade, boats were pulling up to collect people fleeing. Rosen was struck by the absence of panic. With the Hudson's strong current and nowhere to dock properly, captains of the vessels could only jam the prow of their boats against the wall of the esplanade. There were no breaks in the esplanade guard rail for people to board. Two men hoisted up the elderly Rosen and, in her words, "threw" her over to the boat, where two more people caught her.

> I met six other people on that boat from this building, and we just stuck together. We arrived in New Jersey. I don't think we had a dollar among us. We were not welcomed, we were not expected. They did not know we were coming. A woman offered us change and said, "I hope this helps." We hitched a ride in a utility service truck. There were no seats in there, just

coils of wire. Spools. And we were just going to a hotel because one of the ladies accidentally found a credit card [in her pocket] and she got the rest of us in.

When she left her home, Rosen said, she "had no ID, I didn't have a penny. I had nothing else." She added, "I didn't move back into my apartment until the end of October," like most residents, remembering that date clearly. "October 25th."

Depending on how much damage their homes suffered, when their apartment building reopened, and when they felt safe, residents who decided to come back did so several weeks to several months later. Some of those who ultimately returned did not move in until January. They came back to a very changed neighborhood.

The first change was the dramatic alteration of the shape and space of the neighborhood as a result of direct destruction, isolation from the rest of

A poster mounted at Ground Zero during an anniversary of September 11 shows Battery Park City at the moment of the collapse of one of the towers, as smoke and debris engulf Battery Park City buildings. In the foreground, ferries scramble to rescue residents fleeing to the waterfront.

the city, and recovery and reconstruction efforts. The events of September 11 continued to have daily effects as they disrupted everyday activities, circumscribed use of the neighborhood to residents and rescue workers, and provided residents graphic reminders of catastrophe.

Second, residents used the parks, street corners, playgrounds, plazas, and promenades of Battery Park City to reconnect with friends, find people who had shared and could discuss this experience, ease the pain of the events by talking to others, and rebuild their community by developing bonds with other residents. In this way such space was a catalyst not of destruction but regeneration, and demonstrated the value of preserving familiar, shared spaces whenever possible for communities affected by disasters so that they can use these settings to rebuild community ties. Residents used their neighborhood not as public space open to all but as what I call *community space*. In a community space, the people who use it reserve it, unofficially, for use by other members of their group. In Battery Park City's community spaces, members of the community could expect that others shared common experiences and could be talked to. Residents' use of space in this period challenged two sets of claims: first, the claim, in earlier critiques of the neighborhood, that its public spaces were so superficial as to be nonfunctional, and second, the claim in current disaster research that disasters, rather than uniting a community, reverberate long after they occur by destroying the mutual trust necessary for social institutions and tearing communities apart along long-dormant fault lines. Battery Park City residents in fact consistently reported that in the months after September 11 they grew closer to their fellow residents and that more community bonds were formed. Some found it perverse that such a horrible event had had positive effects on their community. While several factors contributed to the postdisaster unification of community, the fact that residents in this disaster could reconnect in their original community spaces demonstrated both the resilience of the design of the neighborhood and the strength of this community. More generally, Battery Park City's experiences demonstrate the need, in all communities, for a range of different types of public space, including community space.

Changes to Public Space

The collapse of the towers and the cleanup that followed affected daily patterns for residents, workers, visitors, and anyone who came to Battery

Park City for a considerable time afterwards. Whereas in the rest of the city disruptions of daily life lasted only for a number of days, in Lower Manhattan (and at the Pentagon in Arlington, Virginia) such disruptions lasted for months or even years. These physical changes were one of the ways the disaster's impact reverberated. At every turn, alterations to residents' daily lives made life in Battery Park City much more difficult.

That Lower Manhattan was directly affected by September 11 much longer than other areas is significant for the disaster's social impact. In *Bowling Alone*, his 2000 study of the decline of civic participation in America, Robert Putnam concludes that a significant cause of higher rates of community involvement by older people was the nearly four years of American participation in World War II. The prolonged, shared sacrifice required by the war trained that generation to participate at higher rates than subsequent cohorts did. Immediately after September 11, commentators expected a sea change in citizen participation. Putnam's own nationwide follow-up survey in late 2001 painted a more modest picture. Attitudes had changed, but participation had not. "Much of the measurable increase in generosity spent itself within a few weeks," he concluded. But Putnam had explicitly argued in his earlier work that World War II's pronounced effect on civic participation was due to its disruptive effects on citizens' everyday lives for an extended period of time. After September 11, only Battery Park City and the surrounding area of Lower Manhattan were disrupted for an extended period. The extent of physical disruption to residents' everyday lives would set the highwater mark for the extent to which September 11 altered community participation across the nation.

Returning residents confronted a dramatically altered landscape not only across West Street at the Trade Center site but just outside their doors. For some time, police and National Guard stationed at barricaded streets allowed only residents with identification to enter Lower Manhattan and only permitted entry into some buildings with an escort, for such purposes as retrieving pets or medication. Once I could make repeated tours of the area, I saw National Guard soldiers and weary recovery workers traversing the streets and paths. The pedestrian bridges to the World Trade Center that had been residents' primary link to the outside world had been destroyed, and for a year afterwards travel outside the neighborhood was complicated by disrupted transit service. The Trade Center subway station and PATH train station were gone. The Crystal Palace–like center of the Winter Garden Mall was shattered and inundated with debris, and some residents later saw rescuers recover human remains concealed beneath the wreckage.[2] Playgrounds, sandboxes, and other outdoor areas were contaminated with asbestos and other toxins from the collapse; eventually all the grass in the parks had to be torn up and replaced before the lawns could be used again. One playground was now within view of Ground Zero. Heavy machinery, emergency equipment, and security patrols converted the esplanade from parkland to a slow-moving thoroughfare, made more challenging by large wooden barricades running down parts of it that covered thick tangles of temporary, above-ground power lines.

Community garden plots were buried under temporary stacks of crushed vehicles pulled from the rubble, and leaking gasoline from the cars and trucks made the garden unusable even after the vehicles were removed. Most of the garden was eventually covered by a replacement pedestrian bridge over West Street, further south than the original bridges. The North Cove, once a centerpiece of the development that boasted extravagant, transoceanic yachts, was cleared of pleasure ships and lay empty. While September 11 remained prominent in people's minds in many places, only in a nearby community like Battery Park City did I observe how residents' most mundane, everyday activities required negotiation of many physical obstacles that insistently brought up memories of the terrorist attacks.

Even after restrictions had been lifted against nonresidents entering the area, people remained cut off from their lives outside. With the loss of the World Trade Center, service was suspended at the nearest train stations. Roads closed for recovery work similarly disrupted bus lines, and West Street's closure for many months made travel by car or taxi more difficult as well. Though planners had never foreseen it, residents had relied on the mall in the base of the Trade Center complex for much of their local shopping and were suddenly without basic services like pharmacies and banks, or places to buy clothes and other goods. Some residents lost their jobs, either because of the destruction of their workplaces or because of the ensuing economic downturn. Others saw their jobs relocated from damaged Financial District buildings to back offices across the river, in Jersey City, or elsewhere. In addition, the smaller number of stores in Battery Park City did not immediately reopen. "We had celebrations when the pharmacy came back. When the bank came back. The last thing to open was a grocery," remembered Eleanor Rosen. She reported that being cut off from important transportation and services had made even common chores difficult to complete. Whereas in many residents' descriptions these practical difficulties were a metaphor for the emotional difficulties of living in the neighborhood, Rosen, as an older person living alone in Battery Park

Four months after September 11, 2001, the waterfront esplanade was obstructed by slow-moving rescue and recovery vehicles and by wooden barricades carrying temporary power lines (*foreground*). The Winter Garden Mall (*center*) was severely damaged by the collapse, as were the office buildings of the World Financial Center. Many windows are still boarded up on the tower to the left.

City, spoke in a qualitatively different way about what had happened. Older residents discussed the practicalities of the situation more readily than the emotional weight of September 11, either because they thought it inappropriate to articulate such emotions or because they had had more traumatic events in their lives—wars, the Holocaust, the death of a spouse—in which to contextualize September 11 than did younger adults, for whom there was no comparison to the devastation of that day. Rosen talked about the pharmacy that wasn't there anymore, not the towers that had been outside her window that were now gone. In detailing the reasons why some families hadn't moved back, Rosen described the closures of schools, not trauma, fear, or concerns about environmental pollution. In those early months, other residents' discussions moved from these physical obstacles to emotionally freighted aspects: how frightening it was to leave, how joyful people were when they came back, or how they felt when they saw a neighbor again for the first time and knew the other person was safe.

Just out the window of some apartments on Chambers Street, cranes dump tons of twisted steel from the Trade Center site into barges destined for a New Jersey recycling firm.

Other changes to the space had a persistent effect on residents that they only realized later. Linda Edwards had just moved to Battery Park City in 2000. Now her apartment looked over where the barges docked to haul away the World Trade Center. Day and night, crews filled the barges with massive, twisted girders and other debris from the buildings.[3] "I didn't re- alize how much that affected me until they were gone," said Edwards. "I looked out of my window and took a nice deep breath. To make it through those months I blocked it out. But when they were really gone, I felt how heavy they had laid on my heart."

Every weekend, moving trucks lined up along the North neighbor- hood streets near where Edwards lived, as more families moved out. Many headed for the suburbs of New Jersey, others for affluent neighborhoods in Brooklyn. Some were concerned about the respiratory health of them- selves or their children, both because the air outside had been filled with pollutants from the collapse and the months of fires, and because when the towers collapsed, Battery Park City apartment buildings had been inun- dated with contaminants like asbestos, lead, and other heavy metals. Other residents moved because their apartments were too damaged for them to move back in, or they felt too anxious or frightened being in Battery Park City, or they were upset by being so close to where people they had known had died. The weekly arrival of the moving trucks added to the anxiety of those who still lived in the neighborhood. Many times, Edwards and her husband asked whether they should move, too. They felt as though half of the residents in their building had left. But her husband was looking for a job, which could have meant they would have to move twice.

Signs of environmental concerns and the possibility of more attacks were evident in public spaces as well. In that first winter, air quality monitoring equipment was installed on poles and lampposts along the streets, a constant reminder of the unknowns in the air residents breathed. But even when the equipment was removed, some residents told me they were unsure whether that meant the danger had passed or whether it would just be ignored.[4] By summertime, I spent many days at the main playground playing with my daughter and occasionally speaking with parents. The space was once again popular with kids, families, and caretakers. But adults would look up from the sandbox whenever aircraft flew low along the Hudson River, on loud but unspecified missions. (The local newspaper similarly described the "uneasiness and flashbacks for wary residents" prompted by heightened security during a presidential visit to New York.)[5] The river had always been a busy lane for air travel, but the addition of military and police helicopters on patrol made every passing flight seem more significant. As in other communities across the country, flags hung from windows. For a time, Gateway Plaza's windows were a sea of flags, but they also hung from places like the uninhabited shell of the abandoned Deutsche Bank building, shrouded in black fabric, which stood just across from the World Trade Center and was visible from parts of the park. Collectively, these visual cues where residents walked every day to work, on errands, or to the park added a constant, low murmur of stress to life in Battery Park City.

Several of the changes altered the influence of the physical neighborhood on residents. West Street in 2002 had a very different effect than it had had before 2001, when most people crossed on the pedestrian bridges, not at street level. Battery Park City's most noticeable barrier went from being easy for businesspeople and residents to cross but difficult for outsiders, to being physically challenging for everyone. Changes also demonstrated the relative importance of various features to residents. In describing the loss of local services, residents made it clear that they used the World Trade

Top: Windows of Gateway Plaza bedecked with U.S. flags, one year after September 11, 2001. *Bottom*: Several signs posted inside a Gateway store. The one on the far right urges residents—of Battery Park City specifically—to display flags. The sign at left announces a candlelight vigil organized especially for residents. The window also reflects a long line of white television trucks, with their satellite dishes pointed to the sky, that were there to cover the anniversary. Residents were irritated by the crush of press and sought to create symbols and rituals for residents only that would allow them to commemorate the event on their own, away from the rest of the city, country, and world.

Empty and shrouded in protective black fabric, the Deutsche Bank building was visible from much of Battery Park City. (A fire in the damaged building during its dismantling killed two firefighters in August 2007.)

Center Mall (with more shops and more everyday needs) much more than the closer Winter Garden Mall in Battery Park City itself.

The destruction of the bridges (and much of West Street, which was closed for months and reopened in a narrower, altered form) threw a significant neighborhood landmark into flux. The separation between the North and South neighborhoods was even more pronounced during the year of cleanup, when travel along West Street was not possible and the waterfront esplanade was obstructed as well.

As disruptive as they were, many of these changes strengthened the boundaries of the neighborhood. Disaster researcher Kathleen J. Tierney has identified the significance of barriers around the Trade Center site that demarcated rescue workers from mere "sightseers and purveyors of disaster kitsch."[6] Though those visiting the site, some of whom had had personal connections to the World Trade Center, and others who were deeply moved by it, would have disagreed with the characterization of themselves as "sightseers," the effect of the barrier for those on the "inside" was to confirm their legitimacy and distinguish between people with something to do in the area from people who had just come to look. Even after Lower Manhattan was reopened to outsiders, few came to the area in the months while the rubble continued burning. The extended restriction on anyone but residents (and rescue workers) entering the neighborhood underscored to residents their shared status as insiders.

Residents also sought to establish discursive barriers between insiders and outsiders, distinguishing themselves by verbal as well as spatial boundaries. They made a point not to say "9/11," which they said sounded like a vulgar abbreviation of a date toward which they felt particularly proprietary. They would say "September 11th" instead. Similarly, they rarely called the Trade Center site "Ground Zero," as it was called elsewhere, even in New York City.

The attacks altered the physical neighborhood of Battery Park City in ways that not only interrupted regular life but altered the structure of the community. Residents' barriers to everyday movement, symbols of September 11, and the physical destruction of parts of their own neighborhood, to say nothing of the site of devastation at the edge of their community, made September 11 an almost constant and unavoidable backdrop to daily life and pushed residents to find others who could relate to that experience. Residents were cut off and oriented more toward each other. The impact of lives lost on September 11 affected people everywhere. But the physical disruption of space in Battery Park City altered the everyday enactment of community. The nature of this disruption, as it happened to befall Battery Park City specifically, had the unexpected effect of facilitating the reconstruction of community.

The Uses of Community Space

Residents explained that when Lower Manhattan had been closed to everyone but residents, they knew that even strangers they saw on their streets had been through what they had, creating a camaraderie that almost invited impromptu conversations. Even after the neighborhood was nominally opened, because routes in and out of the neighborhood were obstructed, the public spaces they traveled through and spent time in were occupied almost exclusively by fellow residents. Residents used their parks, sidewalks, playgrounds, and the esplanade as "community spaces" where they could be among people with similar experiences.

Community space is a distinct subcategory of public space. The key axes along which public spaces are measured are control and exclusion: who rules the space and who can use it.[7] Whereas ideal, popular public space is effectively accessible to all and is controlled by a publicly accountable body, in community space only members of a particular community (whether a residential community or a community of interest) feel comfortable using the space. A close-knit residential block feels homey for people who live there but unusable to outsiders who are stared at as they walk down the street. The sidewalk in front of my neighborhood's motorcycle shop is a regular hangout for a group of men who compare their customized choppers but is not a place where others linger easily. Community space is distinct from popular space on both the axis of control and the axis of exclusion: users are restricted to a narrower group, and the control used to achieve that exclusion is exerted neither by a public agency nor by private owners but by users themselves.[8]

Battery Park City's experiences demonstrates how community space was valuable in helping residents reconnect, recover from trauma, and rebuild community networks. But Battery Park City residents' use of community space also exposed its downside: community space reciprocally encourages an exclusive definition of the neighborhood.

Public space has been idealized as a place of social harmony or of pleasant recreation, or even as the wellspring of democracy.[9] It may be those things, but with two qualifications. First, public space plays many different roles, and there are many types of public space, one being community space. Second, public space is not only a site of social harmony; social conflict is perhaps even more inherently part of public space, and, as the next chapter shows, residents not only united in Battery Park City's community spaces but adopted an exclusive posture toward others who sought to enter their community spaces. Thus attention to particular functions of community space complicates the idealization of public space by recognizing the different roles, some of them exclusive, that public spaces play in urban areas.

People provided consistent accounts of how the sharply altered setting they returned to strengthened public interactions. In residents' accounts, three features of the space helped them. First, the streets and sidewalks allowed residents to interact with each other—to share stories, meet new people, reconnect, greet other returnees, and simply talk over what happened and what was happening—at any time without planning. The second feature was the esplanade, which even more reliably gave residents an audience and a place to go when they wanted to see people they knew or be in public. It was a place, more than the sidewalks, where they could socialize, whether on their way somewhere else or not. Third, residents felt that because the destruction, rescue, and cleanup work in Lower Manhattan had the effect of even more strictly filtering Battery Park City's community spaces during the recovery process, residents were better able to re-form community bonds.

The relief of being back home and finding that neighbors were okay dramatically affected residents. "There was more hugging and kissing than you can imagine," said Eleanor Rosen. Rosalie Joseph, a resident who became active in several community groups after September 11, like many others felt strongly that in a place that had not always had a tightly knit community the bonds among returning neighbors were strengthened. Residents gravitated to shared public spaces to establish those connections. During the summer after the attacks, Rosalie observed,

> When you see your neighbors grieve, and everybody's stunned and in grief and in trauma, I found that when I came back, people were talking in groups. That never happened in Battery Park City unless you knew people. And I would walk down the street, and see all these people talking who didn't know each other. We had an opportunity to create something very special—something we didn't have. It was always very beautiful. But because of this horrible experience, we had the opportunity to create something beautiful here. If you were here that day, if you were home . . . you had a common bond. You couldn't get in or out of this place unless you had special ID, and you lived here. You were cut off. I felt this community shared a bond.

Joseph described a different social relationship in terms of its spatial manifestation. To her the temporary quarantine of the neighborhood created a stronger bond. This took place because there was a physical setting in which residents would frequently cross paths (each time they left home) and visually acknowledge others who they had good reason to believe shared their experiences. Virtually no residents drove to work, and even those who might leave for certain trips by car from their building's garage had more daily excursions, like running errands, or getting to the subway, that would set them on the sidewalk. Because they were walking, casual conversations were easy to start. The importance of community space in creating a strong social bond is highlighted by Joseph's observation that the shared bond was established when only residents were allowed in—not in the most desperate days immediately afterwards, because the neighborhood had been evacuated and residents were spatially dispersed.

While Joseph could see Ground Zero from her apartment and described the shock of coming back to that view, like other residents she focused on its positive aspects in her accounts of this period. Joan Cappellano, who had become friends with Joseph while the two planned community events together, observed an equally significant, and beneficial, change in Battery Park City and again described the transformation in terms of changes in public space. "It was different [before September 11] than it is now. You would see people on the street, and you would say hello . . . but you would just keep on going. But now what it is is more or less a real community. You see someone, you stop, you talk, you listen." Cappellano added, "Now, you'll talk for an hour if you're outside five minutes," an observation several residents made.

Cappellano recounted that she, Rosalie Joseph, and a third acquaintance had been walking outside together one evening when she noted how none had known each other before September 11. This was indicative, to her, of how people had become more approachable. The multiplication of such connections would lead to changes at the personal, political, and community level, as residents like Cappellano and her new friends would organize new community groups and events.

Cappellano said she had been out more during the first summer as well as being more involved with groups in Battery Park City, like the block party planning committee and a group working to restore the damaged St. Joseph's Chapel. "I've had an opportunity to sit out on the esplanade, read, or sit with a friend," said Cappellano. "A concert at Wagner Park, Monday, we had lunch with a friend. And what we all find too is that we're sticking close to the neighborhood." While she said she had more free time the summer after September 11 than the one before, she hadn't been able to explain why she felt so attached to her neighborhood until she was talking with a friend. "And I said, 'It's funny. I really don't like to go out of the neighborhood.' And she said, 'Well, did you ever think that you left once and couldn't get back?' . . . So now, go uptown? Are you crazy? It's really funny," she said with a laugh, more comfortable in the community that understood her best.

While some residents said they had shied away from the esplanade in the spring and summer after September 11, many more said they frequented the space more often. Those residents said it was important that, when their thoughts became preoccupied by September 11, they were feeling down, or they wanted to connect to other people, they could go to the esplanade. Others spent more time in the parks after losing their jobs in the post–September 11 economic contraction. The constant supply of people, the pleasant setting, and multiple excuses to be there meant that the space could be recognized by individual residents as useful for such solace, even though it was not explicitly designated for that purpose. For many residents, the parks and esplanade served as the setting for the ideal open-air support group. Because the area was already a popular park, entering it was a low-investment activity and held none of the stigma that some would have felt about a formal support group. The park was available during far more hours than a support group. Since it was a public space, residents could control the extent of private obligations they developed with other participants, thus making it easier to start a conversation.[10] Residents used it, explicitly, to combat depression and feelings of loneliness and to participate in a forum of informal talk therapy about the trauma they had all experienced.

While living in a neighborhood with shared community space had an overall beneficial effect, shared spaces were also forums for disagreements. Barriers could be created during the same period. One resident, who didn't move back to her home until January, in part because of concerns about pollutants and environmental hazards for her young daughter, found that she couldn't talk to others, even friends, about when she and her family had decided to move back: to those who had moved their family back earlier, her delay implied that they were neglectful parents who had put their kids at risk by returning prematurely. Often, once residents decided it was safe to return to a home or an activity that they had previously been off limits for safety reasons (particularly reasons of air pollution), to continue discussing the risks would be too disruptive. "It puts up barriers. It made me a bit angry. It made me feel like I couldn't talk to the members of my community. My other friend from down there . . . she doesn't really want to think about it or discuss it. A part of me feels angry at her at a deep level." The tension between her and friends would have occurred with or without community space. But interactions in shared spaces could lead to disagreements with others. During the spring, when the refurbished playgrounds and warm weather meant children came out to play in large numbers for the first time since September 11, this parent kept her child out of the playground sandbox out of concern that, although the sand contaminated in the collapse had been replaced, the new sand could have accumulated more contaminants still circulating in the environment. A particularly outgoing woman who was normally quick to make new friends, she was surprised to find herself in conflict with other parents: when others wanted to know why her kid could not go in the sandbox, she became defensive, responding, "I don't ask you why you put your child in the sandbox, so don't ask me why I don't put her in." In negotiating the shared space of the playground, her personal decision to keep her child out of the sandbox became a public statement. While the sandbox was this parent's particular concern, all residents risked conflict over the boundaries they had established in their minds to stay safe or calm: places they would and would not go, routes they preferred to take, concerns they voiced or kept to themselves. Not all disagreements were played out in public space. But because public space is often idealized as a place of harmony and goodwill, it is important to acknowledge that community space makes room for both social harmony and contention.

Though public space has an inherent potential to be a site of conflict, Battery Park City and its temporarily heightened community filtering fostered resident connections to one another, supported residents' return and recovery, and provided built-in opportunities to be in public. Both the already-existing urban quality and the exclusive aspect of Battery Park City contributed to a sense of community space that residents valued in their recovery and the reconstruction of their community. Community space convenient to where survivors are living allows people to have the kinds of supportive public interactions that residents described here.

City of Comrades against a Natural Disaster

Battery Park City residents' experiences in the years after September 11 provide insight into an ongoing debate on the post-traumatic effects of disasters on communities. There have been two different expectations about disasters. The field traditionally held that disasters produce a strengthened sense of community as people pull together to help each other, producing what has variously been called a "democracy of distress," a "community of sufferers," a "post-disaster utopia," an "altruistic community," a "therapeutic community," or a "city of comrades."[11]

For the sociologist Kai Erikson, however, three decades of research on disasters and their effects on communities produced a very different conclusion. "Among the most common findings of research on natural disasters . . . is that a sudden and logically inexplicable wave of good feeling washes over survivors not long after the event itself . . . as people come to realize that the general community is not dead after all," Erikson conceded. But "nothing of the sort happened in *any* of the disaster situations" he studied. In contrast, "these disasters . . . often seem to force open whatever fault lines once ran silently through the structure of the larger community, dividing it into divisive fragments."[12] While there is no consensus on Erikson's findings, the current view is skeptical of traditional expectations of harmony.

Contrary to the emergence of fault lines that fragmented communities Erikson studied, in Battery Park City there was near unanimity on many issues (as will be seen in chapter 7), and on no issue was the community polarized and divided. Not that the potential for such fault lines didn't exist among renters and owners (who could have argued over rental subsidies), among newcomers and old-timers (a difference old-timers acknowledged only with reluctance), among people with children and those without (over park usage, school construction, environmental hazards, noise, or a host of other potentially divisive issues). But those differences did not become determinant.

The community in Battery Park City was not torn apart like those Erikson studied. In contrast, residents reported a strengthened sense of community. Attention to the role of space in the process of recovery suggests an explanation for the very different findings by Erikson and others that challenge the traditional, therapeutic community view. In many of the disasters characterized by conflict, victims either came from different communities (such as the survivors of a sinking cruise ship) and therefore recovered in relative isolation, or were scattered by the very disaster itself: the formative case for Erikson was the Buffalo Creek Flood, in which all the residents of tightly knit communities lost their homes when a mining company dam broke.[13] In that case, the scattering of residents afterwards across dispersed emergency housing sites denied them contact with or proximity to their own neighbors, leaving them among strangers and without the strong sense of community that had developed in the isolated valley where they had lived. "In effect, then, the camps served to stabilize one of the worst forms of disorganization resulting from the disaster by catching people in a moment of extreme dislocation and freezing them there in a kind of holding pattern."[14] Since then, the dislocation of New Orleans residents after Hurricane Katrina has had a similar effect: residents who were still displaced had more difficulty coping and recovering, since "the home is an important asset for coping with adversity, as it is at the core of individuals', families', and communities' rootedness."[15] One report found that much of the secondary effect of Hurricane Katrina "may be attributable to the large-scale population displacement."[16] In contrast, Battery Park City residents returned to their homes and were thereafter surrounded by a community of fellow survivors. A highly restrictive community space provided the forum in which people whom a resident met could be assumed to be comrades. Battery Park City residents' postdisaster experience was in this way more similar to that of the hurricane and tornado survivors studied in the early twentieth-century literature than to the Buffalo Creek Flood: residents recovered in a community of fellow survivors, with the kind of casual public contact more common in well-established communities built on a pedestrian scale than in FEMA trailer parks or autocentric suburbs. The spatial organization of a community is a significant factor in the different outcomes researchers have found when studying disasters.

Another conclusion regarding Battery Park City's response to the disaster can be drawn, cautiously, from the study of recovery in other communities. Erikson noted that "people . . . seem to respect a profound difference between those disasters that can be understood as the work of nature and those that need to be understood as the work of humankind."[17] All of the community-rending disasters he studied were caused by people, not nature: an underground gasoline leak in suburban Colorado, the mercury poisoning of a river running through an Indian reservation, the neighborhood near the nuclear disaster at Three Mile Island.[18] Human-made disasters are so destructive because "the real problem in the long run is that the inhumanity people experience comes to be seen as a natural feature of human life rather than as the bad manners of a particular corporation. They think their eyes are being opened to a larger and profoundly unsettling truth: that human institutions cannot be relied on."[19]

This makes a paradox of the Battery Park City case. The attack on the Trade Center was obviously a human-made disaster, yet residents' unified response was inconsistent with Erikson's findings of division over thirty years of studying human-made disasters. The contradiction may hinge on how Battery Park City residents talked about September 11. Residents did not focus on the people who committed the attacks, did not voice anger at the perpetrators (who were rarely mentioned), and did not generally debate whether the disaster demonstrated that institutions had failed to protect them. Mostly, residents described the unpredictability of the event. They wondered if they would be the victims of another attack, saying, "You never know when something else is going to fall out of the sky." This is not to suggest, of course, that residents were oblivious to actual people's culpability for the attack, nor is it my intention to minimize such responsibility. But the themes of residents' discussions were not recrimination, blame, revenge, or distrust. They did not talk much about whose fault it was or what they thought of those responsible. As throughout much of New York, the primary focus remained sorrow over what had happened, not consideration of human responsibility. (Despite much overlap in emotion after September 11 in New York and the rest of the United States, there were differences of emphasis. The bumper stickers on firefighters' private vehicles parked outside local firehouses said things like "All Gave Some, Some Gave All," or "Never Forget." Virtually unseen were bumper stickers with a picture of Osama Bin Laden and the caption "Wanted: Dead or Alive" that could be found elsewhere. This same contrast was evident in Battery Park City residents' discussions as well.) Thus, when federal hearings concluded that in a range of ways government had failed to prevent the destruction of the Trade Towers, this thread of discussion aroused little interest of any sort in Battery Park City.[20] Residents described much more fear than anger in the wake of September 11: anxiety that such an unpredictable horror could happen again, not anxiety that people might harm them or that their government could not be relied upon to protect them. Residents of Battery Park City did not react like victims of a human-made disaster because they did not frame it as one. Dormant fault lines in the community were not activated, as they were in earlier cases, because residents did not develop the generalized social distrust that victims of other human-made disasters do.

Any consideration of Battery Park City's more successful recovery must include the recognition that residents there had more financial and political power than most victims of disasters do. Residents who lost their jobs after the post–September 11 economic downturn often did not lose their homes because they had sufficient financial resources to get through several months of unemployment. They received rent reductions from landlords and rent subsidies from the government. They could fund community organizations that organized events for the community. They lived in an affluent, state-subsidized enclave whose parks agency could quickly renovate the community spaces after the disaster and plan new social events in the parks specifically to ease children, adults, or the whole community through a traumatic period. Residents had many justified criticisms of the city, state, and federal government's responses. Still, those governments responded more quickly than they have to other disaster-disadvantaged communities, providing direct monetary assistance, reconstruction plans, and other aid. In a range of ways, the community's financial and political power facilitated recovery. Just as researchers regularly stress that disaster "has *social as well as physical* dimensions," so we must recognize that recovery has *physical* as well as social dimensions.[21]

In this respect, Battery Park City is a model for how communities should be provisioned to recover from a disaster. Survivors need easy, casual community space in which they can, as they see fit, interact with others to help them through the recovery period. They need abundant resources from the government, such as the thousands of dollars in rent relief that Lower Manhattan residents obtained, as well as other support for ensuing economic hardships. Finally, residents benefit from the provision of programming and events from agencies like the Parks Department that facilitate strictly voluntary socializing and interaction among community members. In these ways, the re-forming and rebuilding of Battery Park City's community, the resilience of its neighborhood spaces, and the generosity of people and government beyond the boundaries of the community can be a model when traumatic events happen elsewhere.

Space for Community

The physical destruction and disruption in Battery Park City meant that a return to the normal patterns of everyday life took much longer for residents here than in other communities. The shattered windows, blocked streets, and closed stores insistently reminded residents at every turn of an event it was painful for them to think about. Yet despite these constant challenges, residents of Battery Park City described how the neighborhood had aided them in their emotional and collective recovery after September 11.

Consistent with Putnam's findings about the effect of national crises on civic participation, September 11 reshaped community participation more permanently in Battery Park City than Putnam found for the rest of the country. For nearly a year, residents would encounter obstacles to their most basic daily activities that were due to the physical destruction of the surrounding area, and for years afterwards they would talk, organize, lobby, and battle over reconstruction plans. The extended disruption meant that civic involvement in Battery Park City was reshaped by September 11 as it was nowhere else.

Community spaces were important to residents throughout their first year back in Battery Park City as sites where they could reconnect with old acquaintances, develop new ties, and talk through their experiences with fellow survivors. They could also be sites of conflict between residents' opposing opinions. But such places still served as a forum in which some progress could be made toward resolving, or at least accommodating, such disagreements. Residents' use of community spaces in this difficult period demonstrated both the unexpected resilience built into Battery Park City's neighborhood and the particular strength of the community connections residents developed.

The community functions of this kind of space demonstrate a contradiction that must be appreciated in advocating for public spaces: that among its varied functions, public space includes both inclusive, very public activities, and more intimate ones. The latter can lead to dangerous exclusivity but are no less important for communities. While continuing to fight against widely decried tendencies toward the privatization of public space, advocates must also recognize the usefulness of some select spaces that play less public roles.

Residents' experiences, their conception of the attack, and the resources they were able to access in its aftermath set Battery Park City on a very different course from that of other communities that have lived through disasters. But rather than being an outlier, Battery Park City provides a valuable case study of how these spatial, financial, and community resources facilitated residents' reconstruction of a supportive community. At the same time, as the next chapter shows, the resource-rich environment and sense of community that helped heal Battery Park City also fostered resident resistance to the presence of the general public in the neighborhood, and led residents to endorse exclusive proposals for how Downtown redevelopment should proceed. Striking a balance between intimacy and insularity remains a difficult task in disaster recovery, but Battery Park City's experiences highlight how the spatial organization of a community, and its inclusion of community spaces, can lay the groundwork for the kinds of interactions that help survivors cope with the difficult period after such an event.

5

Every Day Is September 11

Memorial Plans for Community Spaces

IN THE YEAR after the Trade Center attacks, Battery Park City residents worked to revitalize their community. While most said they wanted to return to "normal," there were new developments they couldn't ignore. The neighborhood had been thrust in the spotlight. Redevelopment plans for Ground Zero, along with the large number of people making pilgrimages to Lower Manhattan, had the potential to radically alter Battery Park City's separation from the city at large. Residents in a neighborhood that had always enjoyed its reputation as an isolated "suburb in the city" had to choose between restoring that isolation or using redevelopment to reinvigorate the community.

One July night the year after the Trade Center attacks, residents had gathered on the promenade to protest a plan to displace the pleasure boats in the North Cove with a new ferry terminal. Well after dark, the temperature still hung above ninety. The heat was slow to disperse from the marina, as were the residents who had attended the protest. They sat on a long bench near the temporary memorial for emergency workers, watching the water and evening strollers, and talking about their fears for the community as new plans threatened to alter the neighborhood's seclusion. Among those changes were not only the ferry plan but memorial plans for the World Trade Center, a tunnel for West Street, and the revitalization of Battery Park City itself.

One resident described how September 11 had been different in Battery Park City than anywhere else. "Most people [elsewhere] remember it and feel for it, but their experience is over. So whenever we come together— I come to everything [every community event]. It's our way of making a small town in the big city. I think it's becoming like a small town, where people really try to help each other." Residents listed the ways they'd been connecting themselves with the community: there was another Community Board meeting coming up soon, and more meetings against the ferry plan. Rosalie Joseph, seated closest to me, and Joan Cappellano, at the far end of the bench, had been central in planning Battery Park City's first block party. "You'll see all about it in the next *Broadsheet*," they promised.

As Joan and Rosalie described the block party, someone walking by stopped and suggested having a parade. The residents on the bench discussed the other ways they were supporting the neighborhood. Cut off from the rest of the city, residents patronized the few stores that were open out of necessity and loyalty. Joan explained, "A lot of us don't want leave the neighborhood. We support the restaurants no matter how disgusting they are."

Rosalie nodded in resignation. "I haven't eaten at another restaurant, and I'm sick of the food."

But residents' efforts to rebuild their community also left them frustrated at outside disruptions.

Joan shook her head. "I just want the tourists to leave," she said. Looking in the direction of the shuttered firehouse across the street from the Trade Center itself, she explained, "What I want is when it's 95 degrees out, and they're out front of Engine Company Ten, and they're ten people deep—"

"They have a right to be there," said Rosalie Joseph, staring ahead at the promenade, disagreeing with Joan but avoiding a more direct confrontation by not looking her in the eye.

A member of the Community Board sitting between them defended Joan's frustration. "They were not supposed to do that. They had promised not to do that" (to make that sidewalk a viewing area for Ground Zero). "They shouldn't have put that viewing platform there."

Rosalie, whose own work for local groups consistently celebrated the community and sought to make connections among people, appreciated her friends' frustration but resisted joining the Battery Park City consensus that outsiders were intruders. She turned toward me. "I know a lot of people feel that way about the tourists. I feel that it wasn't our tragedy. We don't own it. People should be allowed to get to work, but it's something we share with the whole world. In a way I appreciate them being down here. You see their faces and what they've experienced."

I hesitated, then asked Rosalie if she thought residents' response to these changes had to do with envisioning their neighborhood as an exclusive oasis.

"I think a lot of people are inconvenienced and annoyed by it [tourists visiting]. And understandably, because our lives have been inconvenienced then and since then. We're the backyard. When everybody's talking about community and tolerance"—as they had in the series of post–September 11 public meetings about the future of Lower Manhattan—"we're part of it. We didn't ask for it."

Rosalie's effort to present both sides of the problem tempered the tone of the conversation. Joan tried to restate her position while narrowing her objections. "It's just that one little spot. I'm telling you today, and it was so hot. They were just horded together." She threw up her hands, then let them fall into her lap and shook her head. "I know it's fine. They gotta come down."

Rosalie concluded her assessment of how residents have dealt with the changes. "It's interesting for us because we were so close in history. And sometimes that overwhelms me." Another neighbor explained, "Most people are [living] here because it's a very quiet neighborhood. So it's very ironic."

Three riders on small dirt bikes, all boys, black, not yet teens, zipped by the seated group and wove between pedestrians.[1] They stood out from the other passersby, and there was a pause among the group that Rosalie made a point to fill. She said she was glad life was coming back to Battery Park City. "You know how great it is to see that?" Rosalie said about the boys on bikes. "You know how great it is? It was so depressing before." But Rosalie was alone in her sentiment that the boys represented a return of life to the promenade; none of the other neighbors joined in.

Rosalie had the kind of personal orientation that had led her, in the course of her work in the community, to try to make Battery Park City a more inclusive place. Her welcome of the black boys on bicycles was not typical. Despite Rosalie's vision of the neighborhood as a more inclusive place and her desire to see visitors' needs respected, most residents ended up adhering to Battery Park City's inwardly turned priorities. Influential decision makers used physical and programmatic plans to shape Battery Park City into an exclusive enclave. Elite plans, however, do more than exclude outsiders; they provide the conditions to socially reproduce exclusion, by establishing a residential community that defends neighborhood exclusivity as its very definition.

Residents emerged from their recovery process highly distrustful of proposed new uses of space in the neighborhood. Many different plans for the public spaces of Battery Park City and, of course, the adjacent World Trade Center site were proposed in the years after September 11. Two sets of changes in particular—proposals to install various memorials in the neighborhood and the use of Battery Park City during the annual anniversaries

—brought out residents' anxieties about the changing uses of the community's space and illustrated the central role space continued to play in this period. Plans to set up a memorial or to host anniversary events in Battery Park City provoked widespread opposition and anxiety. Though community space was vitally important to residents, defining a space as being for the community was predicated on excluding others. As a result, residents' desire for community space led them to oppose more popular public space.

Residents' comparative separation from the rest of the city and proprietary attachment to the local parks and open spaces significantly influenced their positions on emerging redevelopment issues. Even before September 11, most residents did not want more people to discover their "best-kept secret in New York." The reciprocal relationship was in operation: Battery Park City's physical design led residents to conceive of their community in an exclusive way, which in turn led them to oppose the new memorial and commemorative uses. Residents found memorials problematic for two reasons. Located in the neighborhood, the memorials would remind residents of an event many were trying not to be confronted with at every turn. Just as important, they would disrupt the symbolic definition of the area as community space by inviting outsiders in.

Memorial Plans

Work in and around Battery Park City proceeded quickly, and by September 2002, residents reported that the physical landscape had largely returned to "normal," which they had said they wanted it to do as soon as possible. But changes to the community meant that in many different ways things could not be as they had been before September 11. In addition to plans that were being made for permanent memorials on the Trade Center site, at least three temporary ones had been installed or planned around Battery Park City. In a climate of great uncertainty over what might be built or rebuilt in the area, residents opposed most of the memorials' presence in Battery Park City and sought to minimize the shadow cast over residents, figuratively speaking, by memorials built on the World Trade Center site.

For the first year, a temporary memorial to police, fire, and medical personnel killed in the attacks was set up in a tent along the esplanade at the edge of the North Cove Marina. The adjacent police memorial that had been there before also became a September 11 memorial site. Later, proposals were made to place the damaged Fritz Koenig sculpture *Sphere*, which

Fritz Koenig's *Sphere* sculpture, recovered from the World Trade Center, was reinstalled as part of a memorial in Battery Park.

had stood in the plaza of the World Trade Center, in Battery Park City as part of a memorial. In the background, as these plans were developed, the selection of plans for the memorial at the Trade Center itself across West Street unfolded slowly and was not finalized until January 2004.

When residents discussed memorial plans, they did not always distinguish between one memorial and another, and a discussion could move back and forth between the temporary rescue worker memorial, Ground Zero, and the *Sphere* sculpture. The consistent tenor of those discussions derived from residents' general opposition to memorials in their neighborhood.

The first impact this opposition had was on Koenig's *Sphere* sculpture. The large brass globe had stood in the plaza of the World Trade Center and was initially assumed to have been destroyed. Not long after the sphere was recovered, relatively intact, at the Trade Center site, downtown redevelopment officials proposed installing it in Battery Park City as a memorial. Residents opposed the plan. As one Gateway Plaza resident explained,

Engravers add the names of officers killed on September 11, 2001, to the police memorial.

"They wanted to put this *Sphere* down right next to this building, which would have inundated us with thousands of tourists. And we fought and fought and fought and eventually it went where it should have gone . . . They have never taken the needs of the community into consideration in making any decision. The only people they invite to these meetings are the 'victims' families,' if you will. So the community is very angry." (That speaker articulated a resentment residents felt toward the victims' families that would only grow as the planning progressed.) Residents felt the sculpture would be too disturbing to see on a regular basis, that it would attract crowds, and that retrieving and reinstalling it was macabre. The plans were changed, and the piece was instead installed in Battery *Park*, the southern tip of Manhattan just below Battery Park City, over which no group took such a proprietary interest.

In contrast, there was little discussion of the police memorial, a permanent installation that had stood in Battery Park City since before September 11. The monument was recessed from the esplanade and included a small plaza with a fountain and a stone wall into which the names of every New York police officer killed in the line of duty were inscribed. Twenty-three names were added for those killed on September 11. Adding that many names was a significant undertaking and was completed just before the first anniversary.

The meaning of the memorial was transformed, from a space that had been more of a curiosity to a casual passerby (Who were those officers killed in the 1800s?) to one of more intense remembrance, particularly for those who knew one of the new names. A year after the attacks, the monument first and foremost memorialized police officers killed on September 11. Residents did not voice opposition to this memorial. The change in meaning and usage that the memorial underwent was subtle.

More concerns were raised by the temporary memorial for emergency service personnel—the 343 firefighters, eight non–New York Fire Department emergency medical technicians, thirty-seven Port Authority police officers, and New York City police officers—lost in the Trade Center. A curved white vinyl tent-style roof was erected along a narrow section of the promenade, next to the permanent police memorial. Under the tent family and friends pinned portraits of police officers and firefighters, departments from across the country left patches and emblems of their own in solidarity, and spouses left personal mementos like flowers and a heartbreaking prenatal sonogram photo.

On the first anniversary of September 11, families, television crews, and others all gather in front of the white tent covering a temporary memorial to emergency service workers near the North Cove.

The temporary memorial consistently drew visitors, sometimes even a small crowd. Residents didn't say that passing it upset them by bringing to the surface thoughts of people who had died, but they were unhappy about the "crowds" it drew. They were also concerned that the memorial would prove not to be temporary at all and would instead remain indefinitely, as some were proposing. Ultimately, officials stuck to their initial plan and removed the memorial soon after the first anniversary.

Residents expected to make use of various memorial spaces but wanted to be able to choose when to enter them. They were wary of plans for memorials in their neighborhood or other places that they would walk by frequently. One resident appreciated that the Battery Park City Authority's greater control over the neighborhood gave them the authority to remove "street memorials" that individuals or groups put up along the sidewalk. These included flowers, posters, and handwritten signs about September 11 and the loss of life. "People were making all kinds of grandiose claims" about the emotional power of informal memorials, she said. "But it's really upsetting passing those things every day when all you want is to live a normal life." Wishing not to once again be confronted daily with reminders of the tragedy they had endured, many residents objected to plans to insert memorials into the spaces they used every day. As they pointed out, memorials are useful to people who make a special trip to see them, but they could be disturbing for residents who had to pass them every time they went out. Residents did not hesitate to consider their own psychological distress ahead of others'. They faced real challenges in their own recovery that they felt the memorials only heightened. But they saw their difficulties as distinct from those suffered, for instance, by the families of both civilians and rescue workers killed in the towers, and they could neither accord space for other sufferers nor find common ground in which to recover.

Residents' concerns went beyond the emotional toll of being reminded. They disliked the fact that the memorials attracted crowds, and they disparaged those visitors as "tourists." To residents the memorials disrupted the more intimate membership they had felt particularly in the months after returning to the neighborhood. The memorials were symbolically and practically popular public spaces, disrupting community space. They were an inherent invitation to outsiders, and outsiders came to use them. Residents objected to the ways new memorials would change the use and meaning of Battery Park City's public spaces.

Another aspect of residents' objections was deeper than, but related to, their definition of space. Particularly in extended meetings about the main

World Trade Center memorial, the priorities articulated by Battery Park City residents consistently minimized the size and visibility of the memorial. Though September 11 was a traumatic event for all residents, neither residents, Port Authority officials, nor others from Downtown insisted in those meetings that the memorial be significant or of an adequate scale to represent what had happened. Speaking at numerous public hearings and comment sessions about the plans, as well as in private conversations, residents expressed more mundane concerns: that the memorial include paths to facilitate their walk to the subway, that it not obstruct their trip on other local errands, and, in some cases, that it not be so prominent and unavoidable as to confront them whether they wanted to see it or not. The fact that residents spoke publicly in favor of a memorial they could walk past but not have to look at, and that would not interrupt daily activities, suggested strongly that many wanted a minimal memorial; far from needing a memorial to help remember the event, they preferred a landscape that didn't force them to think about it any more than they had to.[2] Conveniently, the Port Authority, which had lost many employees and its own Twin Towers, established design guidelines that also encouraged the most minimal monument possible.[3] Part of getting back to "normal" meant trying to think less often about what had happened. Residents also saw no risk of forgetting what had happened and so had no immediate personal need for a memorial. In meetings, residents questioned memorial designs' accommodation of pedestrians, visitor parking, or street circulation, either avoiding or disregarding the much greater emotional stakes the designs had for themselves and others.

Residents were also concerned about visitors to the memorial. "Well, once the memorial is built there will be a tremendous number [of people]," said one resident. But speaking in private, she thought there was little grounds on which to object. "Look, this is public land. The esplanade is beautiful. The marina is beautiful. South Park is beautiful. As much as those of us would like to keep it private, it is city property." In conversations about not wanting more people to come to Battery Park City, it was striking that residents made no allowance for people who were coming to remember loved ones who had died in the towers. While they felt that government redevelopment officials accorded victims' family members excessive consideration, residents tended to make no meaningful distinction between people coming for solemn, deeply important reasons, and warm-weather throngs relaxing in the park on the weekend. Unable to keep perceived outsiders away from their public spaces, residents instead hoped they could minimize memorials that would give anyone else a reason to come to the neighborhood in the first place.

Residents agreed that designers would need to consider the needs of two different audiences for any memorial: people who came to the area specifically to pay their respects to those lost on September 11, and people who passed by several times a day with less emotional preparation, seeking only to get to work or order a slice of pizza down the street. One resident worried about the difficulties a memorial could pose "as we drop our children off at school, carry our clothes to the dry cleaners, and attend the festivals celebrating the rebirth of BPC. Another memorial will be built, a vital memorial for those lost on September 11. . . . [But] every day is September 11 for this community and the contractors and political leaders must remember that when making the decisions that will deeply affect our daily lives."[4] Residents felt strongly that the needs of people who lived and worked downtown were being overlooked, while public attention focused on the desire to grandly memorialize the event and the needs of families who had lost loved ones. Residents' sense that they were being ignored while relatives of the victims received greater attention would produce more explicit tensions as the planning process for the main Trade Center site continued.

Playing host to a continuous memorial procession demanded a great deal of Battery Park City residents. Few communities would take on such a role without members' expressing some frustration after the first six months. (In most other parts of the country, that frustration would probably have focused on the overwhelming problems of traffic and parking that so many visitors would cause almost anywhere but Manhattan, where locals came by train and more distant visitors arrived on chartered buses.) Residents generally did not express what frustrations they sometimes felt to visitors directly. But while such a role would have taxed any community, the expectation that their neighborhood become more open and public contradicted these residents' deeply held, shared sense that isolation had defined Battery Park City. The preference for isolation built into a secluded, dead-end neighborhood—even one built next to the finance capital of the United States—accentuated the tension between residents and the public. Isolation was helpful to residents in the initial period of emotional recovery, but as the public began to trickle back into the area, it made more difficult both residents' recovery and their relations with outside groups invested in the redevelopment and recovery process.

The plans for the major World Trade Center memorial, like the plans for smaller, interim memorials around Battery Park City, belonged to the generation of monuments built since Maya Lin's Vietnam Memorial: stark walls of names carved in humble rather than monumental forms that seek to remember individuals rather than mythologize heroes and events.[5] But the World Trade Center plans faced a challenge unlike any other recent American memorial because of where it would be built. Even the Vietnam Memorial in Washington, D.C., though located in the center of a major city, does not have a residential neighborhood nearby, and people do not pass through the area on everyday errands.[6] Designing a memorial that would satisfy the public's desire to commemorate a major disaster but would simultaneously not offend the sensibilities of locals who had to walk through the area on other business was a significant challenge, and only time will tell if the sunken, below-ground design chosen for the memorial will accomplish this. But residents' sense, as detailed in the next chapter, that their needs were being ignored led them to state their own preferences so insistently that they appeared callous to others' suffering. It would be unfortunate if residents had to walk blocks around a memorial to get a hamburger. But while residents were normally careful to preface their objections with some recognition of victims' families, their arguments in favor of shaping space and constraining memorial design never really acknowledged the scale of what they knew had happened across the street, where nearly three thousand people had died, where mothers had lost their children, where people's long-term coworkers had died in the very offices they had shared; where almost one in ten of the city's firefighters had been killed and thousands of survivors had seen people just like themselves die in unimagined violence; and where, though New Yorkers tended to focus on September 11's local impact, the justifications for two wars had been fabricated and the nation had been reshaped for at least a decade to come. Such momentousness at once both condemned residents' quotidian concerns to seem petty, and excused residents from having to face a memorial to those events on a daily basis. Nonetheless, residents' expectation that the decisions regarding the memorial should be made on the basis of local concerns and not be connected to the discourse of a larger public reflected the aggregation of the uncompromising, zero-sum posturing nearly required in community politics, the prerogatives of socioeconomic and political entitlement, and the insulation of a community defined by physical isolation. It is hard to imagine a community that would have fared well in the spotlight shone after September 11, but Battery Park City's response to the memorial and redevelopment process, despite the neighborhood's advantages in resources, accommodating physical design, and a strong ethos of social service and community involvement, was handicapped by the spatial and social exclusivity that had long served it well. The tensions among residents and others became more acute as the one-year commemoration of September 11 drew near.

The First Anniversary: September 11, 2002

For many residents the one-year anniversary was a magnet for anxiety they had been feeling about Battery Park City's new position. Battery Park City was required to play a very important and very public role. As with the memorials, residents felt anxious, both because the anniversary altered the definition of Battery Park City's parks and streets as community space and because the presence of mourners and others denied them the ability to choose the times and places in which they would think about the event.

Like others, resident Jeff Goldman saw these problems in terms of the challenge they posed both to residents' recovery and to the character of the neighborhood.[7] Residents sometimes described feeling harried by visitors at times when they didn't want to share their own experiences, or simply feeling obstructed by them when they had somewhere to go. Goldman, speaking in his freshly renovated office in the World Financial Center, had been stopped twice by reporters in one week and was frustrated that no one seemed to appreciate that people employed in the area had work to do and meetings to get to. At home a stone's throw from Ground Zero, Eleanor Rosen was similarly concerned about the approaching anniversary. "I'm uneasy about September 11 [2002]. I really don't want to be put through this whole thing again. I don't have to be reminded. I don't want to experience what might be a lot of crowds and disruption, and security. So I might just leave. I would just like to avoid it," she said. Many residents said they were unhappy about the pedestrian traffic created by visitors to the site, even when they recognized that others had a right and a need to come down. These objections reflected residents' frustration over the difficulty of "moving on" when visitors were asking questions about the past, but they were just as much a product of the insulated ethos of Battery Park City. Whether talking about viewing platforms near Ground Zero or more people returning to the waterfront park, residents expressed unhappiness. Most regretted the increased use of Battery Park City's public spaces. Linda Edwards said she and her husband were put off by the increased foot traffic in the area and the neighborhood's increased media exposure, citing

A resident walks past journalists preparing to film during the first anniversary. Some residents said they planned to leave the neighborhood during the anniversary.

her husband's complaint that "New York's best-kept secret is getting more popular." Others worried that the first anniversary would be an example of times to come, when visitors to a permanent Trade Center memorial would overwhelm the neighborhood.

In the days leading up to the anniversary, to be sure, Battery Park City was inundated with television crews and security forces that set up on rooftops, and events for congressional and United Nations dignitaries the previous week had periodically blocked streets and sidewalks for motorcades, delegations, and security. The attention at times could be tiring. But while there was a palpable alteration to Battery Park City on the first anniversary, it was not one defined by impenetrable crowds of people, as some residents had feared.

For most participants, the official commemorations of September 11 structured the day but were only one part of it. Commemoration ceremonies began at dawn. In Battery Park City, elected officials, the head of the Battery Park City Authority, local residents, and others gathered for a modest multifaith candle-lighting service against the backdrop of the Hudson

Left: On the first anniversary of September 11, fierce winds tear at the flags that have been draped on the sides of Battery Park City buildings. *Right*: During the commemoration, a wreath-bearing procession makes its way along West Street, past the line of television satellite trucks. In the foreground, the lawn of the World Financial Center is still dirt after the cleanup operation.

River. Later that day, elementary school children in Battery Park City and neighboring Tribeca started the morning by singing songs together in their schoolyard or telling stories in class.[8] The main ceremonies around Ground Zero began at 8:46 a.m., the time the first plane had struck the North Tower. Family members took turns reading a selection of alphabetized names of those who had died over a sound system that reached the blocks facing the site. Crowds gathered on sidewalks and streets around the Trade Center to listen. At 10:05 and 10:28 a.m., the times the South and North Towers of the Trade Center had collapsed, people stopped while church bells tolled. Afterwards, those who had watched and participated in the ceremony began walking about by themselves, with someone else, or in large groups they had traveled with. They walked around the Trade Center site and Lower Manhattan to be among other people like themselves; to be present, publicly, as part of the anniversary; to find places to reflect on what had happened; and to see other people who were doing the same things. People peered at the site itself from wherever it was visible past construction fences. The day was clear and fresh, not unlike September 11, 2001, except that an unusually heavy wind thundered off the Hudson River, periodically blotting out conversations and tearing fiercely at flags tied to buildings and rooftops. Over the course of their walking, many people, though only a small percentage of everyone who had come to the observances, made their way through Battery Park City.

What was striking about visitors that day were not their numbers, but the purposes and identities they wore quite literally on their sleeves. In addition to New York police and firefighters in formal dress uniforms, firefighters from Yonkers to Japan, Canadian Mounties, and high-hatted British constables came to pay their respects.

Many friends and families of victims attended. Some had small pins or tags that held a photo or name of a loved one. One group wore T-shirts with a picture of a couple who had died. Sitting at an outdoor café eating lunch were two tables of people wearing T-shirts that read, "Together Forever United Flight 175." The presence of that many people who had so visibly lost so much created both a strange sense of public exposure and a sense of communion.

The use of Battery Park City by these visitors did not physically overwhelm the neighborhood in the way residents had expected. At noon, typically the park's busiest weekday time, I conducted timed counts of people

Left: Canadian Mounted Police, interviewed by a journalist in Battery Park City on the first anniversary of September 11. *Right*: Not far from Battery Park City, a journalist and a photographer speak with London police officers. During the first anniversary, police and firefighters from around the world came to pay respects. Numbers of international visitors decreased steadily with each passing anniversary.

entering from each of the waterfront park's eighteen entrances. A comparison to similar counts completed on other days showed more people than usual, but not many more. While a more typical day might produce a count of 135, on September 11, 2002, 168 people entered the park in the two-minute intervals I measured. Much of the esplanade was no more crowded than it would otherwise be at midday.

Abby Ehrlich, who monitored attendance at park events for the Conservancy, made a similar observation when asked if the first anniversary was crowded. "No. We expected many, many more people on September 11 [2002] than we got. There was hardly anybody." Only along the North Cove was there any noticeable increase. The cove was directly west of the Trade Center viewing areas, was home to the temporary memorial to uniformed service personnel and the permanent police memorial, and was already the most crowded location at lunchtime, since workers from the World Financial Center would walk out of their buildings and eat there. Even here a cyclist could easily have biked along the promenade except, at times, on the portion in front of the temporary memorial. There might have been more people, but the space was well suited to accommodate them. Residents had no way to know what to expect on the first anniversary, but the ease with which Battery Park City's open spaces accommodated its share of the day was instructive for future memorial plans.

The day of the first anniversary, the problem was therefore less the number of people than what they represented. Battery Park City was opened up to everyone, with cameras rolling and groups from around the world walking through. It was a space not for the community but for the global public. Similarly, while the events that day didn't have any significant effect on residents' ability to get around the neighborhood, they did affect the experience of passing through it and confronted residents with images of September 11 as they did.

People made subtle alterations to public space that day that may give further indications of how areas may be transformed for memorial purposes. The Irish Hunger Memorial had recently opened and was heavily used that day. But since there is scant indication on the site that it commemorates a hundred-year-old famine, the winding path up a windswept hill that culminates in a view of the Statue of Liberty was easily appropriated by visitors seeking a place to reflect on what had happened a year earlier. It became yet another unofficial memorial to September 11 in Battery Park City.

Alterations to the space also reflected the fact that in public space people adjust their behavior to observe implicit rules and norms.[9] One of the only

The first anniversary brings a small crowd to the temporary memorial to uniformed service workers (*top*), but much of the rest of the park appears no more crowded than usual (*bottom*).

Above: In place of
the Winter Garden's
pedestrian bridge, a
glass wall is shared
by office workers and
visitors as an unofficial
viewing platform over
the Trade Center site.
Right: The renovated
Winter Garden Mall.
In the background,
stairs that once de-
scended from a pedes-
trian bridge from the
World Trade Center.

At the one-year anniversary, a police officer stands in the memorial's courtyard in front of the wall inscribed with names, while others stand at a distance behind a ledge, recognizing an informal demarcation in the public space.

routes from Ground Zero to Battery Park City led directly to the police memorial near the edge of the marina, and a slow stream of people continued to arrive there during the late morning and early afternoon. At the memorial, a two-tiered use of the space was almost universally adopted throughout the day: civilians stood at a slight distance, behind a ledge, to look at the names, while the courtyard below was implicitly reserved for officers themselves. Virtually no civilians walked into the courtyard, though on other days people normally did. At sunset that evening, residents were similarly able to use the more secluded South Cove for a more intimate ceremony organized specifically by and for residents. As William H. Whyte suggests in his studies of successful public spaces, most of these spontaneous changes in use or meaning were successful because people are more self-regulating in their public behavior than we have come to expect.[10] The day presented a useful example of how a well-designed and well-organized space can welcome large numbers of people for such an event without being disruptive to residents or workers.

The day struck a balance between the serenity of the setting and the violence that was being commemorated, between the decorum that sought to provide people privacy and the very public place where their emotions were on display. Around the marina, TV reporters staked out sunny spots on the promenade and conducted short interviews with police officers from abroad, family members of victims, and other attendees. In front of the police memorial, officers from around the world introduced each other, while NYPD officers greeted colleagues. The wind drowned out other people's voices, making the scene more peaceful with its imposed silence. Then a woman in a shirt commemorating one group of victims doubled over and vomited into the bushes, friends grasping her arms and holding up her body as her knees buckled, and the misleading serenity of the day receded before the anguish still fresh in everyone's mind. Such ruptures occurred throughout the day, as someone began sobbing or one family member leaned on another in a weary embrace. Whenever this happened, bystanders' reactions betrayed not embarrassment at a socially awkward moment, but unadorned empathy with the person's difficulty. People came knowing that private emotions would be publicly exhibited that day, though there was little experience to guide protocol in such a situation.

Plenty of people worry about crowds when special events are held in their neighborhoods, and the lack of precedent gave Battery Park City residents added reason to worry that the one-year anniversary ceremonies would make their neighborhood feel chaotic and overrun on a day when some might have particularly wanted solitude. But with the exception of television and security crews, which were concentrated across the street from the World Trade Center site, near the World Financial Center, these fears proved generally unfounded. This was partly because Battery Park City's exclusive design did its job, and most of the people who attended the ceremony never wandered to the waterfront. Contrary to expectations, there was plenty of room for locals and visitors in Battery Park City.

Residents continued cautiously negotiating their new public and private expectations. While the first anniversary indicated what future commemorations would hold for the neighborhood, anxieties did not disappear in subsequent years. Residents' appreciation of the functions of community space reinforced Battery Park City's traditional aversion to public use. Thus even in the absence of overwhelming crowds, residents consistently opposed placing memorials in the community. The neighborhood could accommodate such uses, but they would have been inconsistent with the community mission residents saw for the space.

Reluctant Memorialists

Battery Park City adapted remarkably well to the added demands of hosting ongoing memorial functions after September 11, demonstrating surprising versatility for a planned community. Public space is defined in part by being meaningful to people and capable of being reshaped by them. The fact that numerous spaces around the neighborhood were easily reconceived as memorial spaces made Battery Park City more important to residents and visitors alike. It was a significance that no one wished for, but one that decisively altered the meaning of the neighborhood, from a residential enclave that was the purview of financial-sector elites to a space the public used to observe, enact, and recall the collective memory of September 11. But residents did not view this successful adaptation of the space positively. The neighborhood becoming meaningful to so many people set residents' private conception of Battery Park City at odds with public needs for the space. This conflict would become even more pronounced as more Lower Manhattan development plans were debated.

On ensuing anniversaries, the day continued to be one of significant public use in Battery Park City, but the neighborhood was not overwhelmed, and residents took advantage of more intimate spaces in which to mark the event. It was never an easy day. Memorializing can be important, but it can also be exhausting—it was for me, even several years afterwards, and even though I had not been in New York on the day of the attacks and did not know anyone who worked in the towers. The anniversaries were not normal days; residents continued to treat them differently, whether attending events or just walking down the promenade to be present in the commemoration. But after the first anniversary, few residents talked about leaving the neighborhood for the day. Still, many residents' fears persisted. Those fears were fostered by the same sense of tight-knit community that had helped residents connect to each other in the months immediately following their return. Those ties were part of the reciprocal relationship in which the exclusivity of the community would later lead members to reproduce that exclusivity in future designs for the space.

Battery Park City residents carried a significant responsibility as de facto participants in September 11 commemorations. But no neighborhood was better positioned to do so. Battery Park City was robust. The community was resource-rich in many ways and could even provide ready-made landmarks and memorial spaces. In return, outsiders observed implicit norms about behavior or restrictions on the use of different spaces. Residents have

Behind the World Financial Center, arrays of spotlights near where the Trade Center towers stood in the "Tribute in Light," September 11, 2003.

resisted this exchange with the general public, but they have personal reasons to embrace it. As the number of current residents who were present on September 11, 2001 diminishes with age and as people move away, the community's response on anniversaries will increasingly be established by people with less personal connection than those who ran for their lives or for their children that day. It would be a loss if the reluctance to having visitors crystallized into resentment toward visitors. Current residents are the ones who can ensure that the community respectfully welcomes people who come to remember what residents themselves experienced. The exclusivity of Battery Park City has shaped the community, but it is not its destiny.

Recognition of the contribution of community space can be used to secure what is socially distinctive about Battery Park City. Residents repeatedly expressed uneasiness with their community's parks becoming popular public spaces, and regretted, or even quietly resented, perceived outsiders in "their" parks. Distinguishing community space from popular space could help resolve some of those anxieties. Residents might well be more comfortable with the popular, public role of the waterfront esplanade if they could be assured, by the Battery Park City Authority or by the design of certain spaces, that some other part of the neighborhood would be preserved as community space. One possibility is that parks that are further inland from the esplanade could be designed to serve the community more intimately as a "backyard" to complement the very public front yard of the esplanade. Alternatively, space that is not public, such as building courtyards, swimming pools, and recreation space, could be identified collectively and explicitly as community space. Similarly, programs intended to serve the local community, such as the community center or the annual block party, could be tied together thematically to be places where the community's identity could be strengthened and reproduced. Some of these spaces and events could, alternately, be an opportunity for residents to present their community's identity not only to each other but to the visiting public as well. In these ways, the community could develop a secure sense of itself without defining itself so strongly in distinction to outsiders who also have legitimate claim to public spaces.

Recovery and redevelopment presented important opportunities, but these were not immediately realized. Instead, residents' need for community space and the neighborhood's endorsement of an exclusive definition of community would lead residents to oppose plans to better connect Battery Park City to the rest of New York. As subsequent chapters show, residents responded to redevelopment plans by organizing to use redevelopment to reinforce the neighborhood's isolation.

6

Class and Community Organizations

WHETHER BATTERY PARK City residents had sufficient community organizations depended on how far away an observer stood. From the greatest distance, critics of Battery Park City had long assumed that the "simulated" citadel neighborhood would have only a thin imitation of community life. Residents just outside Battery Park City, however, openly envied how organized the neighborhood looked to them. Speaking at a public viewing of competing plans for the World Trade Center memorial, one Lower Manhattan resident said she "envied" Battery Park City because they were more organized. "All they have to do is put a flier up in one lobby to get the word out to a thousand people." Another Lower Manhattan resident thought Battery Park City residents were *too* organized and was angry that they could use their political clout to mop up minor inconveniences while her block was still facing serious problems. At a community planning workshop, she was frustrated that "the big issue for the Battery Park City people was a dog run. I said, 'Excuse me! We have [the abandoned shell of] Deutsche Bank, lights in our windows, constant noise, crowds, vendors, and two feet of debris we had to shovel out of our homes ourselves. A dog run?'"

Meanwhile, in Battery Park City residents worried about how their neighborhood would suffer for not being organized enough. One resident organizer explained, "The only way Battery Park City is going to be listened to is if it gets organized separately instead of relying on the better organized neighborhoods to the north like Tribeca and SoHo . . . We're sort of the stepchild to them." Another activist agreed. "BPC has never really been a politicized community. It's been mostly a transient community. People come stay a few years, and have kids and move on. So there never was any kind of organized community groups there."

Two distinct influences shaped the community activism in Battery Park City: the physical space of the neighborhood and the socioeconomic class of the residents. This chapter examines the varied way each of these factors shaped a range of community groups before and after September 11. The physical boundaries that defined the social community played obvious roles in the positions groups took, the alliances groups forged, and how groups defined themselves. Class mattered, not just in the money residents could contribute at fund-raising, events, but in terms of political agency and residents' views toward community organization and leadership roles.

This presentation of local organizations provides an opportunity to compare elite community groups to those in working-class neighborhoods. A comparison between the activism of Battery Park City groups and community activism as detailed in several classic ethnographies of low-income communities goes beyond the obvious resource differences to reveal unexpected similarities between rich and poor. Qualities long thought to be characteristic of poor and working-class organizations, including parochialism, suspicion of outside organizers, and personification are just as evident in one of the wealthiest zones of the city. F. Scott Fitzgerald's observation that the rich "are different from you and me" is misremembered today as facile.[1] He went on to say that if he were to write about the rich, "I should have to begin by attacking all the lies that the poor have told about the rich and the rich have told about themselves."[2] Fitzgerald concluded that wealth had oxymoronic effects: that it made one soft but cynical, that it imbued a sense of superiority independent of observed conditions. In Battery Park City, even common assumptions about how the rich are different turn out to be misapprehensions. Wealth makes distinctive marks on the community, but not in all of the expected ways.

The Space and Class of Community Activism

Linda Belfer moved to Gateway Plaza, the first complex in Battery Park City, in 1982. She soon joined the group that would become the Gateway Plaza Tenants Association (GPTA). Space and class each gave Gateway residents a distinct sense of themselves and their organization. Gateway was the six-building complex built in accordance with Oscar Newman's notion of defensible space, so not only did the residents share a single landlord, but they shared a single entrance to the grounds, as well as a pioneering sensibility as the first residents in Battery Park City.

Because Gateway was first rented as subsidized, middle-income housing, residents confided that as Battery Park City grew they felt distinct from the greater affluence of the rest of the neighborhood. They were proud of Gateway's comparatively populist profile. The cohesion of tenants and the GPTA was only reinforced each time people banded together to fight off the landlords' periodic attempts to take the building out of rent regulation to charge higher rates. When the Lefrak Organization, Gateway's management company, first tried to raise rents substantially in the late eighties, GPTA successfully opposed that effort, and as resident Eleanor Rosen explained, "Once you have a common purpose, there's sort of a cohesive quality that develops." GPTA has fought rent hikes since then (most recently in mid-2004), and this has maintained the group, residents' awareness of it, and resident activism.

Because of its efforts to raise residents' rents, the Lefrak Organization has often been involved in community disputes. Other management companies, like Rockrose, have occasionally been the focus of attention as well, but they were never more so than in the months after September 11, when residents sought rebates and rent reductions for their uninhabitable apartments, and organized many new tenant organizations to win more consistent concessions from the management companies.[3] Soon most buildings had tenants' associations, but none as well established as the GPTA.

Gateway's distinctiveness influenced the rest of the community in recognizable ways. While a few other buildings also had tenants' associations before September 11, GPTA was central. It represented the largest and oldest complex in Battery Park City. At the time, its 1,712 apartments made up 38 percent of Battery Park City's 4,452 units and housed 36 percent of its population. And Gateway was a wellspring of an even greater share of the community's activists. Gateway was the largest and most middle-class complex in Battery Park City, and it contributed a disproportionate amount of the community's middle-class activism before and after September 11, from dog owners' associations and groups that collected Christmas gifts for children in need to environmental activists and organizers of neighborhood meetings.[4] Residents supported the claim by Belfer, who was president of the GPTA in the years after September 11, that Gateway's "history of activism" established its residents as leading activists in Battery Park City.

The Gateway complex was a center of community activity. It was home to the *Battery Park City Broadsheet* and its founders. Multiple members of Community Board 1 lived there. And the number of groups there grew after September 11.

The Community Newspaper

The *Battery Park City Broadsheet*, published out of Gateway by Alison and Robert Simko, two longtime Gateway residents, was another important neighborhood institution. After September 11, residents consistently cited the *Broadsheet* as one of the primary ways they got information about their neighborhood, and it appeared to have played this role since its creation in October 1997. Rosen brought it up unprompted as a "wonderful" paper that "gives us all the local news." Although other newspapers (like the *Tribeca Tribune* and the *Downtown Express*) covered Battery Park City as part of Lower Manhattan and had broader distribution and far more pages, the *Broadsheet*, which was available almost exclusively in Battery Park City apartment lobbies, covered Battery Park City issues more consistently and was referred to more often by residents.

The paper was indeed wonderful. Well written, it thoroughly covered community events big and small—from the opening of a new local restaurant to the construction of high-rise office buildings and, of course, a major terrorist attack.[5] It was printed on thick, glossy, oversized broadsheet paper much more substantial than even that of the high-end financial newspapers that targeted Wall Street.

Critical for the paper's success was the presence of sufficient consumer demand, in a community of only eight thousand people, to lure enough advertisers to support a paper that came out every two weeks. Few neighborhoods could boast the buying power to sustain such a venture, and the fact that Battery Park City could gave the community an invaluable resource.

The *Broadsheet* helped connect Battery Park City groups by keeping each informed of what the other was doing. The Community Board meetings were part of its regular beat. Its coverage included issues like the battles between tenants and management over rent increases in Gateway, and there was regularly coverage of public space issues, as parks were designed and redesigned and developers and the Battery Park City Authority (BPCA) came before the Community Board for support of their plans. In this way, the *Broadsheet* helped connect the organizations and their members.

Considerable overlap existed among community groups. Belfer was president of the GPTA and a long-term member of Community Board 1. Tammy Meltzer was on the board of the GPTA, the WTC Residents Coalition, and another group. The Simkos lived in Gateway and published the paper from an office there. Additionally, since the paper regularly reported on Community Board 1 meetings and other events, most of the active residents had known of each other's activities for some time. While this is not an unusual situation in local politics, the stable and strong geographic definition of Battery Park City reinforced this tendency: the same boundaries defined the coverage of the local paper, the community board's subcommittee, and the state Authority's purview, as well as a host of other organizations' self-defined area of concern. The neighborhood was defined with the same boundaries for every issue, so players from adjoining areas were rarely added to the equation. Grassroots groups that were connected to the network of organizations in Battery Park City were locally focused, were concentrated in Gateway Plaza, and shared multiple connections among long-term residents. As the field of community groups enlarged rapidly after September 11, 2001, many of these patterns would remain.

After September 11, the *Broadsheet* maintained the focus it had before, but the scale of changes under way in and around Battery Park City greatly expanded what fell under its mandate. Few neighborhood newspapers have had their own photographer capture something like the second plane crashing into the World Trade Center, as Richard Simko did for the *Broadsheet* (see Appendix A). From then on, the paper extensively covered recovery and redevelopment projects, and Alison Simko, at the time its editor and primary writer, attended public hearings of the Lower Manhattan Development Corporation (LMDC), the Port Authority, and the Department of Transportation as well as the traditional Community Board meetings. The *Broadsheet*, like residents there generally, implicitly defined the community (and its audience) as the residents, not those who worked in Battery Park City. It provided little coverage of some events related to workers in Battery Park City: the arrest of dozens of people in the World Financial Center on charges of insider trading got little coverage. The arrest of a resident for walking a dog in violation of Parks Department rules got more. *Broadsheet* stories included coverage of the new, post–September 11 groups, including an ongoing survey by Battery Park City United, new tenants' associations, and public comments by activists at hearings held by state agencies. It remained a center of community information as the field of groups expanded.

The *Broadsheet* did more than report local news, however. Media, whether global or local, do not just report the news, they frame the news. The *Broadsheet's* framing was subtle but distinctive. The paper had a generally "positive" spin on the community, like most community papers large and small. It was not simply boosterish, however. It was ready to cover serious criticism of state agencies, community conflicts, and political issues, and even some disagreements within the community. It avoided, however, strong editorializing. Even the paper's "News & Comment" column was news, not commentary. The section announced new development in the area: the new pedestrian bridge, the founding of a new community group, or new funding for post–September 11 children's health. The paper was not a vent for residents' most base and petty frustrations. Other public venues in the community showed there was potential for more venom. E-mails from community groups expressed greater anger, online discussion boards included more hostility toward outsiders, disagreements at community meetings could be contentious or petty. The *Broadsheet* achieved a surprising balance that created a stronger sense of community while denying that sense of community an exclusive, snobbish, or cliquish tone. The *Broadsheet* even sought to orient the neighborhood toward a set of rhythms different from those that predominated in Lower Manhattan. In an area that rose and fell with stock prices, interest rates, and bull or bear markets, the *Broadsheet* quietly highlighted natural cycles: it was published on the full and new moons, it reported cruise ships' scheduled arrivals and departures from New York Harbor, and it was decorated with photos of the changing weather and seasonal events on the city's waterfront. It was not simply a middle-class paper but, more specifically, a beacon of the cosmopolitan-oriented fraction of the middle class, the segment that other observers of socioeconomic class describe as promoting global peace, racial tolerance, and environmental awareness.[6] In a community otherwise dedicated to financial capital, the *Broadsheet* recognized the strong definition of community that residents held, while quietly presenting a relatively less exclusive, more tolerant and cosmopolitan vision of the community. This orientation affected the neighborhood, particularly in comparison to the influence that more heated and exclusive forums would have exercised in its absence.

Post–September 11 Community Groups

After September 11, activists from across Battery Park City agreed that "people are becoming activists who never were before." At least some of the founding members of every local group—the Coalition to Save West Street, Battery Park City United, WTC Residents Coalition—had never been involved in this type of organization, and established groups like the Community Board saw unprecedented numbers of people attending meetings and seeking to join.

Most of the new groups were charismatic organizations. This was true in the sense that the group was founded by a single individual or at most a small group on which the larger group depended for its momentum. Furthermore, the views of the groups and of these charismatic leaders were often indistinguishable. While the groups were not "letterhead" organizations with no members, most described themselves as having a small core of informally organized leaders with a much larger base of members who received e-mails about developments and events. From this larger base, a small but significant fraction would heed one of the group's e-mails and take actions like writing to elected officials, decision makers, or newspapers, or attending a public hearing. The uncertain membership and involvement made the groups' actual size and representativeness difficult to assess, but this is true of any political body. It turned out that the positions of these groups did reflect the majority positions expressed in my interactions and interviews with Battery Park City residents and in those professional telephone surveys that were conducted.[7] Thus the groups represented most clearly the positions of their founding members, but they also reflected the views of their e-mail-receiving members and a large number of Battery Park City residents more generally.

One of the first groups established after September 11, Battery Park City United, formed to address new issues affecting the neighborhood. Its name clearly framed the discussion and points of view in relationship to a spatially defined neighborhood and set of interests, even though the issues it first addressed, like expanding ferry service into the North Cove, submerging West Street in a tunnel, or considering government environmental cleanup proposals, could have been framed in other ways as well.

Manohar Kanuri, a Battery Park City United founder and Gateway Plaza resident, explained that he and other residents started the group after they had worked together "on different issues that kept cropping up since September 11" but that they had not been involved before that. For his part, Kanuri had never been in a group like this but was a member of Greenpeace and similar membership organizations.

The group was first active on the ground but soon moved to the Internet. Initially, they collected 880 signatures in recognition of the buildings staff throughout Battery Park City, to whom many were extremely grateful for their commitment to residents' well-being on September 11. Maintenance staff had cleaned up, kept buildings operating, and looked in on pets when residents were not allowed back to their homes, and they had showed endurance during the extended period when the neighborhood was without regular services. The signatures were collected building by building. But, said Kanuri, "As time has gone by we've relied increasingly on e-mail directly to our membership and indirectly through a bunch of other lists," including a public school PTA e-mail list, e-mail lists of condo owners, and others.

Despite its shift to e-mail contacts, the group remained highly concentrated geographically. In an online survey conducted on the group's Web site, 89 percent of the sixty-three respondents who provided addresses lived in the south neighborhood of Battery Park City, with only 8 percent from the North neighborhood. More than half were from Gateway alone. Only 3 percent were from outside Battery Park City. This localized membership points not only to the ongoing geographic concentration, despite the expected "despatializing" effects of the Internet and extensive use of e-mail lists as organizing and communications tools, but to the central role of Gateway Plaza as well.

Battery Park City United was an example of a group that had a cohesive identity through residents' strong definition of Battery Park City. But interactions with groups outside Battery Park City ultimately contributed to its undoing. Kanuri had been pushing for the group to ally itself with a group called September Eleventh Families for Peaceful Tomorrows, in an explicit effort to bridge some of the divides between them and other groups they had confronted in Community Board meetings and other settings. (The latter group was a peace organization that opposed the wars resulting from September 11 and was started by family members of people who had died in the Trade Center.) But around the same time, Dave Stanke, another member of Battery Park City United, wrote a piece in the *Daily News* highly critical of some victims' families' expectations for the World Trade Center memorial. Stanke's editorial expressed the impatience many active Battery Park City residents had begun voicing toward family members after long months of meetings about memorial plans. In the piece, he wrote, "The families of victims of the WTC attack already have received far more than other survivors usually get. At some point, they may finally be able to give up their lobbying and deal directly with the only real issue they face: the loss of a loved one."[8] To some members of Battery Park City United, Stanke's piece sounded insensitive, unnecessarily caricaturing family members of victims as selfish lobbyists. Amid these discordant views toward family members of the victims, Battery Park City United fell apart. To a significant degree, its splits came about because members were unable to agree on what the organization's relationships should be with people and groups outside Battery Park City. Battery Park City's spatial cohesion provided both the basis for group unity and the source of its conflicts with groups outside that boundary. It was a breakpoint many of the groups faced.

Elsewhere in the neighborhood, buildings without tenants' or residents' associations started them, and other tenant groups that had been much less central to people's lives found their meetings well attended. In northern buildings like Tribeca Pointe and others managed by the firm Rockrose, these came about during a rent strike over secretive and unequal rent discounts that different buildings were offering to entice tenants to stay when so many were moving out after September 11.

Amid this increase in tenants' organizations, Battery Park City resident Sudhir Jain sought to organize these individual-building groups into the Lower Manhattan Tenants Coalition. From the start, the group (which was renamed the WTC [World Trade Center] Residents Coalition when its membership grew to include not only tenants but associations of co-op and condo owners as well) was an exception among Battery Park City organizing. It was one of the only new groups in Battery Park City not to be defined by Battery Park City.

This wider focus had concrete effects on the positions the group took, most pronounced during debates over the location of a tour bus parking garage planned for the Trade Center memorial (discussed in detail in the next chapter). At public meetings, Battery Park City residents were otherwise universal in wanting the garage built outside Battery Park City. Jain's position, however, was affected by the broader geographical representation of his group; he said he hoped that no single neighborhood would be saddled with all the exhaust and traffic of a large garage and that instead a plan would be adopted that spread the burden around in smaller parking facilities so no single residential area suffered too much. WTC Residents Coalition's initial step of including non–Battery Park City members led it to consider development issues differently than most other groups formed in Battery Park City, because it had to consider how development would affect other neighborhoods, too.

Despite its exceptionalism in not conforming to the boundaries of Battery Park City, the WTC Residents Coalition fit quite well into the network of Battery Park City–centric groups. Battery Park City United's Manohar Kanuri attended early meetings, as did members of the Community Board. Rosalie Joseph made a presentation to the coalition about the Battery Park City Block Party, and Jain regularly attended the public hearings of the LMDC and Department of Transportation, as well as the larger meetings of Community Board 1.

There was another difference between other Battery Park City groups and WTC Residents Coalition in their relationship to other groups. Jain and Tammy Meltzer, another member of the group, were both on one of the LMDC's advisory councils. Other more locally defined Battery Park City groups were unable to even get meetings with the LMDC, much less gain a seat on their advisory boards. Since membership numbers and the representativeness of all these groups were difficult to gauge, it is unlikely that the WTC Residents Coalition was selected because it represented more constituents than the other groups. Rather, it formed very early, and its name (then referring to Lower Manhattan tenants) no doubt made it seem appropriate for an advisory council that was to speak for Lower Manhattan residents. Finally, compared to Save West Street or Battery Park City United, WTC Residents Coalition did not take positions as publicly. WTC Residents Coalition was more cautious both because of the personality of its leader and because its broader collection of groups across Lower Manhattan meant it could not adopt as strong a position as more unified, Battery Park City–centered groups. This moderation could have contributed both to their appearance of neutral representation and to a less oppositional relationship with the LMDC.

Building a Block Party

Battery Park City's spatial boundaries were more than physical; they also held inherently political meaning for residents. Battery Park City's "first annual" block party was very much a product of its neighborhood. It was born when residents who had turned out for the reopening of the local Embassy Suites Hotel were sitting together in front of the hotel. Someone proposed a block party, and a group of residents, who hadn't known each other before, "just decided to run with it." The annual event, first held in

A poster advertising the first Battery Park City block party, September 2002.

late September 2002, had several purposes. It was, certainly, a chance for neighbors to come out together and to reinforce a sense of a shared community. "Help Us Celebrate 'The Best Small Town in the Big Apple,'" urged a flyer for the event.

Symbolically, the block party was always held on a weekend close to the anniversary of September 11. Organizers drew vendors from businesses and residents in Battery Park City both to cement this community identity and, as a second goal, to help local businesses, many of which were still struggling a year later—"to bring the people together, so the local people can make some money," as Joan Cappellano said. Finally, Rosalie Joseph, who chaired the Battery Park City Block Party Committee, also saw a value for organizing the community: she anticipated that residents would become better connected and more ready to collectively represent their community's interests. Joseph's description of the event flowed seamlessly from a sense of community to one of political organization. "This is the acknowledgment of our unshakable spirit. . . . Through it all what has emerged is the strongest sense of community we have ever experienced. Planners have to understand that this community is very much alive, very much unified, and as the plans roll out and construction begins we want to be recognized as a strong voice."[9] Residents' shared trauma made the casual, friendly interactions between strangers that can take place in public spaces easier and more common. These connections and openness developed into a small organization, a community-wide event, and aspirations for greater community involvement in future public decisions.

The block party was also distinctive in its scale. In other neighborhoods of New York, a block party is typically just that: a block long, closed to car traffic for the day, with some hotdogs and hamburgers donated by a local politician, an inflatable ride from the borough president, and neighbors painting kids' faces while people chat with each other. On larger avenues in Manhattan, events like this may be larger but are more generic: a more professional collection of merchants selling anything from jewelry to socks makes the rounds of such events, setting up shop on a different street every weekend during the warm months. Battery Park City's block party was far, far larger than other truly "neighborhood" block parties but also had none of the generic merchants found at most larger ones. Even at the second annual party, booths were assigned only to local businesses and community groups. A mobile stage was brought in, and live entertainment (mostly residents) played all day.

By the time of the second block party in 2003, the committee that organized the block parties had merged with Battery Park City Parents Association (whose existence predated 2001) to form Battery Park City Neighbors and Parents Association. This provided extra people to work on the block party but also gave that group a mission for the rest of the year, which could help them become the broader community mobilization Joseph had hoped they would be. The Neighbors and Parents Association demonstrated some of the straightforward material advantages that socioeconomic class provides Battery Park City groups. One June shortly after the two groups joined forces, they planned a happy hour as a fund-raiser and social event. People came to the Gatehouse, a newly opened restaurant near Gateway, to drink and socialize. Battery Park City residents had no trouble spending money to meet at the bar. Some had paid for child care. In front of the restaurant it was almost too crowded to walk.

The crowd included a mix of ages. There were people in their twenties who looked much like the young, single new arrivals who had come to Battery Park City after September when rents were discounted. Residents, who had "adopted" the local fire station, had invited the firefighters from Ten House, who were there as well. An older generation, including Community Board members and others who were centrally involved in community events, was also well represented. Other people represented groups that were just forming or had not been involved in rebuilding issues, including Manhattan Youth, and CERT, a community-level version of the Federal Emergency Management Agency.[10] CERT, which had organized a neighborhood emergency response team and trained residents in CPR and emergency procedures, had quickly become a popular group in a neighborhood with distinct memories of September 11.

The attendees were overwhelmingly white. The last names of many in the group indicated that they were the descendents of Italian and Jewish immigrants from the wave of immigration a hundred years earlier. But in a city that was nearly 40 percent foreign born, few were immigrants themselves. I commented to Rosalie Joseph that I had noticed more African Americans volunteering at the last block party than at most Battery Park City events. She did not disagree. It seemed to me that events Joseph was involved with tended to be more diverse than most in Battery Park City. But she offered no explanation as to why that had been the case. Joseph, a casting director for television, was naturally gifted at identifying people's talents and getting them involved. A Gateway resident, she, like the publishers of the *Broadsheet*, oriented her community political involvement toward a generally liberal, inclusive vision.

Participants at the fund-raiser passed the hat and quickly filled it. Someone announced to the group an upcoming fund-raiser: for $75 a person, residents could get a ferry boat ride from Lower Manhattan to an upcoming Yankees game. It sounded like a great way to go to a baseball game. A community group that could sell tickets at that price had an advantage over most in fund-raising. But the neighborhood also benefited from an unusually large number of residents ready, willing, and experienced in taking leadership and executive roles. At the happy hour, for instance, Martha Gallo commented that she hadn't been very involved in the community before September 11. But afterwards she not only got involved with the Battery Park City Neighbors and Parents Association, but soon became the president and had been very involved in the community since. Residents took on roles of officers and leaders of the groups with enthusiasm. In this way the advantages that class provided came not only in the form of the $75 fund-raiser tickets but in the types of managerial skills many residents had developed in their work and community lives.

The Community Board: Forum and Advocate

Newly formed community groups often turned to local governmental bodies and agencies seeking to influence plans for Lower Manhattan's redevelopment. Many groups turned first to the local Community Board. The community boards had some elements of community groups themselves. Though community boards in New York City officially have only an advisory role, they are the most local level of government. Their public meetings provide an open floor for residents' comments. They most often deal with planning issues and residents' quality-of-life complaints, and the city looks for the local Community Board's endorsement before approving things like liquor licenses and construction. Members are appointed in equal numbers by the local City Council member or members and the borough president. Because of their connection to local elected officials, community boards are able to draw political support for local issues. (On the other hand, council members expect endorsements of their pet projects from the board members they appoint.) Community Board 1's Lower Manhattan district physically included not just Battery Park City and the World Trade Center site but also City Hall and other influential offices like those of State Assembly speaker Sheldon Silver (who, along with the governor and the Senate majority leader, was one of the three most powerful elected officials in the state). The Community Board was both a significant local actor in its own right, taking on community issues, and a forum where local groups focused their energies and attempted to make their case. While attendance at such meetings waxed and waned as issues came and went, Community Board 1 became more central to Battery Park City after September 11.

Originally, Battery Park City residents were drafted onto the Community Board's Battery Park City Subcommittee almost before they moved in. At first, the Community Board had no residents on the Battery Park City Subcommittee because until Gateway opened there were no residents in Battery Park City. So, as Belfer recounts the story, soon after she joined the tenants' group, "without even asking my permission she [city councilperson Miriam Friedlander] appointed me to the Community Board." (Members of community boards typically apply for a spot. They then serve on any number of the subcommittees that meet each month and attend the monthly general meeting.)

As residents moved into Battery Park City, Community Board 1 became more engaged with their issues. It was also recognized by residents as a place to bring community issues. In this process, it became integrated into the small web of organizations serving Battery Park City.

After September 11, Community Board 1 continued to be a hybrid of a community group and a forum for other such groups. To the previous collection of responsibilities like endorsing or opposing new business applications and zoning variances was added the much larger, far-reaching, and complex involvement in the Trade Center redevelopment going on within the community board's geographical boundaries. All of the major decision-making agencies made presentations to Community Board 1 more than once: the LMDC, the Port Authority, the Department of Transportation, the BPCA, the Battery Park City Parks Conservancy, and still others like Daniel Libeskind, nominally the designer of the Trade Center's master plan. Representatives of all of the grassroots groups attended meetings and spoke with board members. Among other things, this involvement has made the Community Board much more significant since September 11. As longtime member Linda Belfer said, "Certainly, more members of the public are showing up at the meetings. And more people made applications to join the board than ever. Before it used to be difficult to recruit people for the board. Now people walk around disappointed that they weren't chosen this year."

Community Board 1's dual role as a forum and an organization made it both a place where these other groups came to oppose development decision makers directly and a group that these same activists lobbied for support. The conflicts that resulted from this second role meant that local groups were at times quite dissatisfied with the Community Board.

One member of Save West Street complained that "the worst reception we've gotten from any place has been from certain members of Community Board 1, particularly people who live in Tribeca and think they know better than we do what's best for us." This opinion, shared by at least some other activists, reflected in part the effect of West Street's isolation on resident activists.

One irony of straddling the definition of community group and governmental body is that many members of Community Board 1 felt shut out of redevelopment decisions just as other Battery Park City groups did. At the very least, members of the Battery Park City Subcommittee (generally Battery Park City residents) were unanimous in their frustration and sense of being ignored—both as residents and as the local Community Board, from whom they felt the decision-making agencies should have sought more input. One member did suggest that Battery Park City was not alone in feeling ignored and that it was a problem for all of Lower Manhattan. Community Board 1's unpopular positions were blamed on leaders' geographic distance from Battery Park City, while the Community Board's Battery Park City members felt left out because of their own spatial position.

Interactions with State Redevelopment Agencies

One of the most significant additions to the field of local government actors was the creation of the LMDC, announced by New York governor George Pataki in November 2001.[11] Like the BPCA, the LMDC was a state-chartered corporation. Traditionally, such corporations have been formed to perform tasks while insulating both elected officials and the corporation itself from voter pressure and public scrutiny. (They could also be considered more "political" because they were less accountable. Governor Pataki appointed John C. Whitehead head of the LMDC; Whitehead had contributed over $175,000 in the previous election year, including a single $100,000 gift to Pataki's party. A deputy secretary of state under Reagan, he was politically reliable.)[12] The LMDC was a subsidiary of the Empire State Development Corporation,[13] and it grew to have fourteen board members (seven appointed by the governor, four by then-mayor Rudolph Giuliani, and three by Mayor Michael Bloomberg). The LMDC also created a number of advisory boards, presented as a means by which the LMDC could gain input from relevant parties, such as victims' families, Downtown residents and businesses, and cultural institutions.[14]

While the LMDC was officially the planner and decision maker regarding Lower Manhattan redevelopment, the Port Authority of New York and New Jersey (the board of this state-chartered corporation is appointed by the governors of the two states) owned the Trade Center land. They were therefore also involved in the planning and decision-making process. The New York State Department of Transportation was responsible for, among other things, the proposed West Street tunnel along the eastern border of Battery Park City, which became a major redevelopment issue for Battery Park City residents. While each of the three organizations was structured somewhat differently, residents' interactions with them were similar and best discussed together: all three were state-level bureaucracies, all three had the power and money to implement their redevelopment plans, and all three were involved in a series of required public hearings, the most prominent of which were part of environmental impact statement review processes. An environmental impact statement is required for any large project, and as part of the assessment of how the project will affect the built and natural surroundings, the developing agency is required to hold public hearings, solicit public comments, and respond to them in writing. (It is not required to change its plans in response to them, however, only to show that the comments have been considered.)

Community groups and residents found the LMDC unresponsive, but because it was charged with overseeing the planning and construction of the World Trade Center site and related projects it remained a focus of significant amounts of activism. Members of the Community Board, Battery Park City United, the WTC Residents Coalition, and others organized resident input to the LMDC public meetings for the environmental impact statement. Other agencies held similar meetings, such as the meetings of the state's Department of Transportation for the West Street Tunnel. Community groups responded similarly to these.

Such public meetings had the virtue of appearing open and democratic and even required the LMDC to respond to resident concerns. But since the LMDC was not required to act on them, residents felt frustrated by the process even while they continued organizing for similar meetings. Their frustration was born of several features of the process found both in LMDC public hearings and in those of other groups.

At a typical hearing, two or three officials would sit at a table on the stage of a large auditorium. One by one, people in the audience would step to the microphone and make their comments. Not infrequently, residents would ask questions—sometimes actual questions, sometimes rhetorical questions, sometimes angry questions. "Many have been searching for information about toxins they may have been exposed to" after the towers' collapse, noted an environmental lawyer at one hearing for the environmental impact statement. "What can we expect during construction of the World Trade Center?" Others asked if plans to monitor air quality during construction included a way for the public to know what the testing had revealed. The officials at the table would meet the questions with blank stares. If the audience member pressed them, repeating the question, those on the dais explained that they were not there to answer questions during the hearings and were only to collect public comment. A tape recorder on stage would have served the same purpose, but having real people there gave the impression of a responsive organization.

The LMDC or other agency would also limit the relevance of the public comment process by defining the scope of the hearing as narrowly as possible. They would claim, for instance, that plans from years ago had already addressed an issue or that one component of the plan fell under another agency's auspices. That way, elements that residents might object to were not part of the hearing, relieving the LMDC or comparable agency from the responsibility of responding. Thus the Coalition to Save West Street complained that while "as reported . . . in the NY Times, the majority of the comments received by the LMDC on any issue were tunnel objections . . . on this issue the agency did a fantastic job of passing the buck." (Many of the letters had been organized by the GPTA.) According to the LMDC's final statement, "The local elected officials, Community Board 1, Battery Park City residents and others all oppose this tunnel." (The LMDC then named over 160 people who had submitted objections to the tunnel.) But despite this public and very democratic voicing of views, the LMDC report concluded that "LMDC is not responsible for the proposed Route 9A Reconstruction Project; which would be undertaken and is being reviewed by NYSDOT [the state's Department of Transportation]." No response was therefore required. This claim, however, was less self-evident than it might appear. Significant parts of the LMDC plan for the World Trade Center site did depend on the tunnel. The LMDC might have had to comment on the tunnel, therefore, except that designs the LMDC submitted for review included two different sets of plans, one with a tunnel, one without. Then the LMDC could insist that "all impacts of the Proposed Action were properly disclosed" while having to take "no position as to which option for Route 9A is preferable."[15]

Perhaps more frustrating to some residents was their own conclusion that factions of the public were being used to obstruct each other so that the LMDC and other development interests could move forward. Battery Park City resident Manohar Kanuri of Battery Park City United complained that residents' and victims' families' divergent views had been used by the LMDC to divert attention from the LMDC itself.

One of the reasons that the Community Board 1 meeting [with the LMDC, residents, and victims' families] was such a collection of fireworks was because a similar discussion hadn't taken place before. That's what happens when you have a body like the LMDC taking you aside and talking to you and taking me aside and talking to me . . . and I certainly think they used the residents to counter the families . . . They say, "Hey, we threw the idea out there, and hey, the residents opposed it. We do empathize with your desire for a memorial." . . . Cynical, maybe.

The LMDC used the "input without influence" model in other ways. Less than a year after the attacks, it organized a "Listening to the City" program in which the public at large was asked to attend meetings or participate in online forums to express their views on redevelopment and a memorial. Again, Kanuri was dismissive, calling it "a big farce." People felt like "okay, we participated, we said stuff, we pressed buttons, they were tabulated, these are the motions of democracy," recalled Kanuri. But he was skeptical of the process. It was not clear at all that input was used in redevelopment decisions. "Was it a town hall meeting? They had to go through this exercise of public participation. A way to legitimatize this illegitimacy" —by which Kanuri was referring to the fact that an unelected group like the LMDC was making decisions about redevelopment.

Kanuri and others had similar comments about the Port Authority, which as owner of the Trade Center site worked with the LMDC on issues like the reconstruction of a "temporary" PATH train station that they acknowledged would actually be permanent: "I don't think it's a coincidence that the Port Authority 'actively seeks the input of all parties involved.' How politically correct, at such a late stage. They could have started this process long ago. You don't come back and ask for input after you built the temporary station, so I think this is a way to divert the anger of the families away from themselves [the Port Authority]."

Residents were further frustrated by the LMDC and the Port Authority because community groups could not get an audience with them. John Dellaportas of the Coalition to Save West Street said neither organization would meet with his group. They met with even more aggressive opposition from the Department of Transportation: "The Coalition to Save West Street learned this week that its invitation to make a presentation at an upcoming April 21, 2004 meeting of the Civic Alliance—a prominent coalition of more than 75 business, community and environmental groups providing a broad 'umbrella' for civic planning and advocacy efforts in support of Downtown rebuilding—had been rescinded, at the insistence of the New York State Department of Transportation, which refused to present its tunnel scheme to the Civic Alliance if we were allowed into the room."[16] As a result of public outcry, the Department of Transportation soon apologized, and at the last minute the coalition was reinvited to the meeting.[17]

While the coalition could not get a meeting with any of the groups, they could, and did, comment at the environmental impact statement public hearings. In this respect, the hearings were useful to the agencies in at least one way: whether or not they provided information that altered the plans, they served as a focus for activism long after the groups would otherwise have been frustrated with their lack of access and potentially would have tried other more confrontational tactics.

One of the surprising features of community mobilization on Downtown redevelopment was how little energy was focused on decision makers. This is not because groups had not tried or because they were unaware of the players. Rather, the field was cluttered with an array of different yet similar groups with overlapping responsibilities, different levels of visibility, and different levels of popularity. In addition to the agencies discussed in this chapter (the LMDC, the Port Authority, the state's Department of Transportation), there were many more potential targets for public pressure: the governor, the mayor, Larry Silverstein (who held the Trade Center lease and had already begun rebuilding the 7 World Trade Center building), the BPCA (which was involved with proposals that reached Battery Park City's borders), city agencies like the New York *City* Department of Transportation and others that regulated construction noise, environmental conditions, and zoning. In addition, there were people and organizations that community groups sought to influence because they could be valuable allies: newspapers for which members wrote columns and letters; elected officials at the city, state, and federal level; other community groups that might endorse the groups' position (like Community Board 1 or the chamber-of-commerce-like Civic Alliance); and not least of all citizens in the community themselves. The result was that community groups, some with a quite small core of active members, organized efforts to sway the position of an extremely wide array of entities. The benefit for any of those entities, of course, was that pressure on any one of them was dispersed and that in fact some, like elected representatives, might be able to voice support to their constituents without expecting it to have any effect on the redevelopment project.

Government and corporate groups from "inside" Battery Park City like the BPCA fared considerably better than the others in avoiding pressure from community groups. When both the BPCA and Brookfield performed "passing the buck" motions comparable to those of the LMDC, activists' responses were milder. This may have been because they were less centrally involved or because the BPCA had recently developed a reputation among Community Board members for being more responsive to residents.

For instance, people on the Community Board questioned representatives from both the BPCA and Brookfield Properties (which managed the World Financial Center) about poor maintenance of pedestrian bridges, an issue about which the Community Board had become increasingly angry. While board members were not happy that both groups had said, essentially, that the broken elevators and poor maintenance were someone else's problem, members saw it as a tangle of responsibilities that needed to be straightened out, not as an organization that was intentionally ducking resident concerns. This was an unexpectedly patient response, given that all residents present agreed that the condition and maintenance of bridge elevators and access had been unacceptable for years. It helped that both groups were from Battery Park City. Both had productive ongoing relationships with the board. And unlike the other entities like the LMDC, both had agreed to meet with the Community Board.

Brookfield escaped even more scrutiny. When the Coalition to Save West Street was unable to meet with the LMDC, it was able to set up a meeting with Brookfield. As the company that profited from all the office and shopping mall space on that side of the proposed tunnel, Brookfield had good reason to support, or even promote, a tunnel that it felt would bring more traffic to their buildings. One member of the coalition claimed that "the main lobbyist [against the community group's position], to tell you the truth, is Brookfield Properties." But when I interviewed him after the meeting, a coalition member who had met with Brookfield was relatively sanguine about them: he offered a long and reasonable-sounding explanation of why Brookfield's position was in opposition to the coalition's, then concluded, "We came out of that meeting with an understanding about Brookfield's position." It was surprising that more energy wasn't directed toward the organization that residents saw as the main proponent of the development plan they opposed. But discussions about Brookfield had a tone of familiarity and intimacy, not the sense of distance required to vilify a group.

The governmental agencies that became important after September 11 had several features in common, including a set of public input meetings that offered the form but not substance of democratic participation. While residents were frustrated by this, they continued to attend and generally accord legitimacy to these meetings and put less pressure on these groups than they probably would have if they had been denied any meeting at all. Government and corporate organizations from within Battery Park City fared better still, often being considered reasonable or given the benefit of the doubt.

Comparison with Working-Class Communities

Battery Park City's experiences demonstrate the complex ways in which class shapes community organizing. This study of an elite community allows some valuable comparisons to the more numerous studies of poor and working-class communities with regard to community organizing. For the purposes of comparison, I have drawn on a collection of ethnographies of community organizing, including several that, as studies of communities facing their destruction, bear particular similarities to the challenges faced in Battery Park City. The earliest study, *Urban Villagers*, is a classic examination of a working-class, primarily Italian American neighborhood in the West End of Boston. In 1957, the population of the West End was seven thousand, nearly the same size as Battery Park City.[18] Plans were drawn up to demolish the neighborhood and build luxury high-rises in its place as part of the era's "urban renewal." Eventually, the residents' homes were indeed bulldozed. This study provides the basis of my comparison because it looked so carefully at the role class played in the community's response to the threat of neighborhood redevelopment. *Norman Street* is set in New York during the 1970s, when closing firehouses, building abandonment, and other signs of the city's fiscal crisis threatened to depopulate a poor- and working-class, predominantly Polish American neighborhood. Other studies, including *American Project*, on a high-rise public housing complex that was demolished, and *Black Corona*, which examined New York City community organizations, provide useful points of comparison as well.[19]

Each closely examines how the community organized in response to a threat to their neighborhood.

A comparison of working-class and upper-class communities shows that a surprising number of unflattering characteristics previously presumed to be characteristic of working-class groups and values are just as present here. For instance, working-class residents in Boston's West End were often suspicious of organizers and organizations from outside their communities. A civic-minded, upper-class man from outside the community began a "committee" and tried to involve residents in opposition to redevelopment plans but failed to engage them.[20] Gans concluded that "one of the major obstacles to the Committee's effectiveness in its own neighborhood was its outside leadership." Residents had heard of the group's leader, but "they knew also that he lived outside the area, and that however strong his sympathy, he was in class, ethnic background, and culture not one of their own."[21] In part, the aversion to outsiders could be attributed to class differences between often college-educated reforming activists and working-class residents. But an inward orientation and parochialism appeared to play a role as well. And while suspicion of outside organizers could reasonably be attributed to the peer group society that characterized working-class life in the West End, this aversion appeared in upper-class Battery Park City, too. Groups in Battery Park City with outsider organizers were unable to attract residents, while locally based groups became central players.

A local environmental group's experiences serve as a typical example. Environmental issues seemed like a natural issue on which to organize residents. They were a concern to a good number of Battery Park City residents. Some already belonged to environmental groups. Others were worried about the long-term health risks from the dust and debris of the towers' collapse. Still others became concerned about new environmental concerns: some groups worried about exhaust from the bus garage proposed for Battery Park City, and a group of parents were very worried about diesel exhaust when a busy ferry terminal was relocated next to the neighborhood's major playground, which was always full of children. (The ferry terminal was located near the playground partly as a response to resident opposition to locating it in the North Cove Marina.) Yet when an environmental group made presentations (to the WTC Residents Coalition, for instance) they seemed to generate as much hostility as sympathy among residents. Organizing, mobilizing, and communicating are inevitably challenges for community groups. Kim Flynn of the WTC Emergency Environmental Coalition later said the group had been active in other blocks immediately surrounding Ground Zero but had been unable to mobilize Battery Park City residents. The founder of the environmental coalition lived in Independence Plaza, a complex much like Gateway Plaza, little more than a block away from Battery Park City.[22] But Battery Park City residents and community organizations had few interactions with the 1,329 households in Independence Plaza, and this local group never overcame its "outsider" status.

The fact that Flynn had been able to organize residents elsewhere in Lower Manhattan but not in Battery Park City suggests several possibilities about how widespread such suspicions of outsiders are. Different communities vary in their receptivity to outside organizers, depending on the neighborhood's past experience and the community's degree of cohesion. Space appears to play a role. Several outside activists complained about difficulties particular to Battery Park City because of its spatial organization and architectural design. Battery Park City's lack of common elements used by activists—there was virtually nowhere in the whole neighborhood to post flyers—made communicating without an established network more difficult. Likewise, Battery Park City's strong sense of itself made outsiders seem less relevant to this already "institutionally complete" community.[23] Physical and social isolation, rather than class alone, influence a community's receptiveness to outsiders.

On environmental issues, residents took their concerns to other groups based in Battery Park City: the Coalition to Save West Street publicized fears about diesel exhaust from the bus garage. Parents worried about the ferry organized loosely among themselves and brought their concerns to the Community Board.

The environmental group's experiences were not an isolated event. R. Dot (a group whose name stood for Rebuild Downtown Our Town) generated attention elsewhere in Lower Manhattan but not in Battery Park City.[24] Likewise, Battery Park City residents Linda Belfer and Pearl Scher had long been involved—independently of each other—in local Democratic Party committees. But the party did not become a public player in the redevelopment debates, even though many elected officials came and went during the redevelopment period.

Although many communities rejected outsider organizers, doing so hurt Battery Park City less. The community had abundant members of their own community ready to take on leadership roles. Despite the similar reception nonresidents received in Battery Park City and Boston's West End, Battery Park City residents were not oriented toward the peer group in the way West Enders were: their social circle was less local than West Enders. Among West Enders, joining a community group would have been a betrayal of their loyalty to the peer group. In contrast, joining an organization gave Battery Park City residents additional legitimacy in their community.

In an upper-class community such as this one, there was a greater sense of entitlement to take on and lead such work. Herbert Gans wrote that the culture of the working-class West End mitigated strongly against a resident being an effective organizer.[25] Here residents were enthusiastic organizers and gained status by doing so. Many residents were already of the management class, so leadership roles felt natural to them. With such skills and motivation, people sprang up to start organizations, take positions, and talk to neighbors about their involvement. In the year after September 11, it sometimes appeared that residents, when confronted with a problem in their neighborhood, were more ready to start their own organization than to find and join an existing one. For many, class norms weighed in favor of community involvement, not against it.

One of the reasons observers have commented on the parochialism of working-class communities is that such a perspective stands in direct contrast to the cosmopolitan attitude of people who become researchers themselves. Such parochialism seems very different from the openness that upper-middle class researchers imagine generally characterizes their own class. But while field researchers from the liberal professions may invest a great deal in their identity as cosmopolitan and welcoming, it is not clear how determinant that tendency is among the upper class and upper middle class in general. Instead, studies of fractions within larger social classes consistently indicate that classes contain both more cosmopolitan, tolerant, multicultural elements (among the upper middle classes, often correlating with those in the "helping" professions and education, and those with a larger investment in education as social capital), and more conservative, particularistic elements more ethnocentrically interested in their own culture, community, and immediate material interests (and more aligned with those whose status is derived more directly from financial capital).[26] Battery Park City (particularly Gateway Plaza) had many members of the more cosmopolitan fraction, particularly among those who had jobs in television production, journalism, and education rather than finance. But plenty of residents, even among activists, protected their community before the larger world and their financial interests before inclusive political ideals. From this perspective, it is not surprising that some upper-class communities were more oriented toward their fellow members than toward an abstract public good, or that such a community was found in the Financial District.

One difference between Battery Park City and West Enders was that the latter "had considerable difficulty in understanding the complicated parade of preliminary and final approvals, or the tortuous process by which the plans moved back and forth between the Housing Authority, the City Council, the Mayor, the State Housing Board, and the federal Housing and Home Finance Agency."[27] Working-class Boston residents mistook the multiple stages of such a complex process as bureaucratic redundancy and a sign that the project would never actually break ground. In contrast, Battery Park City activists consistently represented the process to their constituents as a linear one, which could either move inexorably closer to a damaging outcome or be slowed incrementally at each step by citizen input. One benefit that class provided was the training to take on massive, complex bureaucracies with confidence. The work that upper-income, white-collar people do prepares them to deal with the bureaucratic communication that is part of the urban development process in a different way than other people might. In the spring of 1958, the residents in Boston's working-class West End received registered letters announcing that they would all be evicted, the land would be taken, and their homes would be bulldozed. Despite the watershed importance of those letters, Gans "doubted seriously" that West End residents read those letters through to the end.[28]

In contrast, when the state generated one-thousand-page-long environmental impact statements, Battery Park City community groups read every word. Groups posted Web links to the environmental impact statements and even complained when these were not readily available in their entirety. Residents on the Community Board produced highly detailed critiques of each report by dividing the thousand pages among four or five members and reading it over in great detail for shortcomings and inadequacies. The group then included those shortcomings in their Community Board resolutions. While the public hearings and environmental impact statements for multiple different but overlapping projects were confusing, other community groups in Battery Park City followed them carefully as well, alerting residents to each new hearing and publicizing shortcomings they found in each draft environmental impact statement. If anything, the groups may have followed the process too closely, since much energy was focused on getting residents to speak at public hearings, even while group members, planners, and other observers claimed that the hearings were typically designed to be only charades of public participation and that the input from such meetings was recorded but not acted upon.

Their ability to persevere through such thickets of bureaucracy notwithstanding, another characteristic this upper-income community shared with working-class communities was the tendency to personify bureaucratic processes. Mayors and governors were often targeted for expressions of discontent as symbols of a much larger process. West Enders did not see the bulldozing of their neighborhood as an inexorable, bureaucratic process but as a decision in which the mayor could personally and definitively intercede. Like West Enders, Battery Park City residents often believed that the motivation behind development decisions came not from faceless bureaucracies but from a single elected official (often, in this case, the governor). This sense was reinforced by times the World Trade Center redevelopment process hit impasses that were widely reported to have been broken only through the governor's personal intervention and as a result of his personal desire to meet symbolic development deadlines (like the setting of a cornerstone for the redeveloped Trade Center on the Fourth of July, 2004).[29]

A related handicap for West Enders—that residents simply "found it hard to think far ahead"—did not appear to have hindered Battery Park City residents. It was difficult for West Enders to imagine what would happen if their neighborhood was torn down, so they more readily assumed it wasn't really going to happen. Likewise, residents in the Chicago housing projects slated for demolition in *American Project* had difficulty envisioning positive outcomes from their displacement because inadequate funding and callous city disregard provided almost no possibility that poor residents could find adequate housing.[30] In contrast, Battery Park City activists imagined in detail what life would be like during the five to ten years that many of the individual projects were expected to take. In this case, the difference between the disruption confronting West Enders and Chicago public housing residents on the one hand—the destruction of their neighborhood—and Battery Park City residents on the other—major disruption and construction on the *edge* of their neighborhood—is significant, but their different class positions are significant, too. Poorer residents quite reasonably couldn't imagine what was going to happen to them after their homes were bulldozed and their neighborhoods destroyed. They had little agency to affect what would happen and few resources to help themselves, so they could not envision any positive resolution. Without minimizing what Battery Park City residents faced, it was less devastating, and the predominantly upper- and upper-middle-class residents had considerably more choices in dealing with it. It was productive for them to discuss in detail the impact to the neighborhood because they sensed they had the agency either to alter the plans (halting the projects) or to make alternative plans for themselves (moving), neither of which was easily conceivable to West End residents. In these respects, both the physical and the programmatic formation of Battery Park City shaped the response of community groups there.

The difference that class-based agency and a habitus of empowerment make was brought home to me in my own Brooklyn neighborhood. My street is composed mainly of two- and three-story rowhouses and apartment buildings. At the end of the street, a block-long, twelve-story luxury apartment tower was under construction. I had been spending weeks looking at how Battery Park City residents responded in detailed fashion to each proposed change to their neighborhood. Residents there came to Community Board meetings to lobby developers to alter their plans. Occasionally projects were stopped, and in other cases designs were modified. Amenities for the community, like libraries and schools, were included in the new buildings. Heading home past the construction site, I stopped to talk to two older working-class women on my block who reliably spent the evenings sitting on lawn chairs in their front yard socializing with neighbors. I asked one woman what she thought of the tower rising in front of her. It was difficult to extract an opinion. "What are you going to do?" she said of the project. When I pressed further, it was clear that she didn't mean that she was resigned to the project but that she was withholding any opinion on it at all. After all, she wasn't in a position to change it. Her opinion had not been solicited before, and she didn't see the point in discussing it now. Newer, more affluent residents had been considerably vocal for some time by then. Likewise, Battery Park City residents, even those who were not involved in community groups and did not attend Community Board meetings, did not believe they were powerless to influence changes in their community.

Thus, at first glance, class differences between working-class and upperclass communities seemed predictable. There were differences in wealth and income, and cultural differences as well. But on closer inspection some differences evaporated while others became more pronounced. Such duality is analogous to that found in opinions regarding the appropriate uses of parks and other public spaces. Since Olmstead and Vaux designed Central

Park, it has been axiomatic that that the bourgeoisie prefer "passive" recreation, such as flower gardens, tree-lined vistas, and promenades, spaces good only for genteel strolls, whereas the working class are believed to be avid users of "active" recreation space for sports and games. In this vein, middle-class preservationists objected when working-class immigrants played baseball in Central Park, whose lawns the designers had envisioned as passive spaces.[31] Indeed, Battery Park City's master plans showed almost exclusively passive spaces, promising "restful retreats from the fast pace of the nearby City."[32] Although the plan promised an "active and varied set of . . . amenities," it provided examples only of passive ones: "lunchtime sunning areas, an esplanade for strolling along the river's edge, large gathering places for public events, and small, quiet interior courts."[33] Later, the BPCA tried to finesse the difference: "Passive space is, in fact, actively used," offered a 1984 annual report.[34] Among the designers' sketches of lawns, gardens, and tree-lined allées were no plans for a single basketball court or athletic field. It should have been a paradise for the upper class.

But residents have since confounded expectations of passive upper-class leisure. At every opportunity, they have argued strenuously for more active space. First, they got involved in the design of the north park and were successful in introducing basketball courts, volleyball and handball courts, playgrounds, fountains for children to play in, and a dock for a community sailing school.[35] They organized to demand baseball diamonds on a vacant lot and fought to make them permanent. During the post–September 11 redevelopment period, they were vigilant to make sure playground space was not lost, dog runs were installed, and community gardens were replaced. Battery Park City members of the Community Board voted unanimously in favor of "more active use" in designs for greenspace renovations in their community.[36] More passive designs were consistently opposed. Thus the expected active-versus-passive dichotomy in working-class and genteel leisure did not apply. But at the same time, the public spaces of Battery Park City clearly belonged to an upper-class community. They consistently had the latest playground equipment, a scrupulously maintained set of baseball diamonds, organic gardening, roof-mounted solar deflectors to bring more sunlight to playgrounds surrounded by high-rises, and public sculpture by well-known artists. The games the children played, the snacks parents brought their children, the activities that people engaged in all reflected the community's social class. Battery Park City's spaces contested long-held assumptions about class-based cultural differences but still exhibited distinct ways in which socioeconomic class shaped the space and behavior of local users.

Likewise, comparing the attitudes and community organization of working-class and upper-class neighborhoods revealed unexpected similarities and important differences. Residents in Battery Park City, like those in working-class communities elsewhere, offered a cool reception to outsiders who sought to introduce new groups and organize residents. Similarly, in the course of a protracted urban redevelopment battle, residents personified the array of government actors aligned against them, attributing ultimate authority to the governor. But beyond these similarities, there were important differences, based not simply in the neighborhood's culture but in an accurate appraisal of the kinds of agency, wealth, and options each community had. Rather than adopt a passive or fatalistic attitude toward large-scale redevelopment projects, Battery Park City residents organized with the expectation that they had the agency and power to alter the plans. While many communities are reluctant to embrace outside organizers, this turned out to be much less of a handicap in Battery Park City because of the number of indigenous residents ready to take on leadership roles and the ease with which new residents could gain acceptance as insiders. Residents had the training, expertise, and class habitus to found and mobilize organizations and group members. In these ways, the class privilege of a community like Battery Park City provides more than additional material resources but compensates for some of a community's less advantageous tendencies. In addition, upper-class communities bring greater levels of human capital, political influence, and a set of shared, class-based assumptions about members' ability to intervene in the plans of government agencies and private developers.

Community and the Spatial Organization of Resources

While the groups most successful in Battery Park City were defined by Battery Park City's boundaries and often tied to the neighborhood by their names, this also meant that the groups did not attract large numbers of people from *outside*, in the surrounding neighborhoods, even when they were facing the same redevelopment issues. A minimal number of people who lived on other sides of the Trade Center site were active in some of the Battery Park City groups. Far more Lower Manhattan residents who would be affected by Trade Center redevelopment decisions were not connected to these groups.

Though there were many new groups, active members were still almost all pre–September 11 residents. Contacts mattered, as well as investment in and connection to the neighborhood. The fact that groups used e-mail lists to contact and inform people (as opposed to posters and flyers, for instance) favored residents who had been around longer. One activist believed that their e-mail lists "cover a significant portion of Battery Park City. Especially the pre–September 11 residents. I don't know how many of the new ones." An effective network of e-mail lists and a disinclination to use more public means to publicize events was biased to involve those who were already well connected over newcomers.

In contrast to almost any other neighborhood in Manhattan, there was almost nowhere to mount posters or broadcast information to the public at large. Whereas bus shelters, phone booths, and telephone poles elsewhere in the city are covered with anything from commercial advertisements to notices of stoop sales to announcements of community meetings or rallies and protests, these fixtures were either absent from Battery Park City or regularly stripped clean by maintenance workers. The only posters regularly visible in the neighborhood were those announcing Parks Conservancy events, which could be displayed in the Conservancy's own glass-covered cases.

Activism in Battery Park City was shaped by the landscape in other ways, most notably by the physical separation of the North and South residential neighborhoods. While the esplanade joined the two halves, the World Financial Center severed the neighborhood's streets and sidewalks. As a result, most residents said their friendships and contacts were concentrated in their own half of Battery Park City.

As important as it is to recognize that Battery Park City's design separated the area from the rest of the city, it was still undeniably urban in comparison to suburban neighborhoods. Because the streets and spaces of Battery Park City were designed as a variation on the model offered by neighborhoods like the Upper West Side, residents could make use of at least some of the unique advantages of an urban street. Thus after a meeting of the Lower Manhattan Tenants Coalition ended with a heated and contentious exchange about how to proceed on environmental matters, a half hour later two of the protagonists and several of the meeting's leaders were standing on the corner under a streetlight a block from the meeting, working things out in an extended, informal conversation. The group could have been filmed for a remake of urbanist William H. Whyte's classic film of New York street life.[37]

Just as the neighborhood's space shaped groups, Battery Park City was influenced by the greater resources the community could muster. These included not only better schools, better parks, better-maintained buildings, and higher-quality housing, but resources community groups could draw on directly, such as the disposable income spent at a block party; the expertise of lawyers, planners, and other professionals who lived there; and the proximity to the Community Board, the State Assembly speaker's office, and other arms of city and state government. These differences in resources gave Battery Park City residents more agency and more choices. Battery Park City residents organized under less loaded circumstances than the deindustrialized, working-class residents of Greenpoint, for whom, "when protest surfaced, it was a manifestation of anger and frustration built up in relation to poor employment conditions, government agencies that caused delay and humiliation, absentee landlords, and inadequate city services."[38] Battery Park City's mobilization was literal opposition to the proximate perceived threat of redevelopment itself, not a symbol of a much larger accumulation of indignities and grievances that had long had no outlet.

The space and class of Battery Park City affected the formation of the numerous community groups that developed there. The following chapter looks at the next phase of the reciprocal relationship between space and the community: once groups were influenced by the physical design of the neighborhood, they organized to shape the physical design themselves. With the economic class and social meaning of the community already etched in its spatial organization, residents sought to influence the redevelopment of the physical neighborhood to preserve existing social privilege and their understanding of community.

7

Definitely in My Backyard

Welcome Nuisances

BATTERY PARK CITY residents not only addressed changes going on within their neighborhood but brought their conception of the community, shaped by the physical design of the space, to bear on redevelopment projects that affected their connection to the city. After the destruction of the World Trade Center, government, nonprofit, and private interests began developing reconstruction plans and holding public forums to propose new designs for Lower Manhattan. Though Battery Park City residents were interested in all of the aspects of redevelopment that would go on across the street from them at the Trade Center site, they mobilized in response to those that could affect West Street, the physical barrier that defined their relation to the rest of the city.

This chapter examines the third stage in the ongoing reciprocal relationship between space and social relations in Battery Park City. In the first stage, people shaped space, as elites planned an upper-class community whose residents would help reproduce the privilege of the Financial District. In the second stage, space shaped social relations, as the space cemented for residents an exclusive conception of what their community should be and influenced their views about nonresidents visiting the area. In the third stage, the cycle came full circle. People shaped space once more, as those residents, informed by the exclusivity embodied in their neighborhood's narrow socioeconomic composition and relative spatial isolation, organized to promote redevelopment plans for Battery Park City, the World Trade Center site, and Lower Manhattan that would reproduce the exclusivity of their neighborhood and community in physical form. Residents' use of space as a tool to maintain exclusivity is particularly pronounced in the redevelopment battles examined in this chapter.

Residents' responses to two components of the redevelopment plan illustrate this third stage of the reciprocal relationship of spatial-social feedback. The first component of the plan that residents organized against in large numbers was the replacement of part of West Street with a tunnel. City and state officials justified the project as needed to better connect Battery Park City to the rest of the city. Residents opposed it for fear it would do just that. The ensuing debate drew out the significant but complex relationship between the built environment and residents' positions on neighborhood issues. The second redevelopment component involved the location of a parking garage to accommodate World Trade Center tour buses and became an emotionally volatile issue. First the Port Authority proposed building the garage on the footprints of the two towers; later it proposed a site in Battery Park City. When the bus depot was to be on the Trade Center site, physically closer to residents' homes but on the other side of the West Street barrier, residents supported it. Once it had been moved further away but onto their side of West Street, they opposed it. In both debates, the barrier function of West Street strongly influenced residents' opinions of what should be done.

As residents moved into the public arena to voice opposition to the tunnel and garage plans, the reciprocal cycle of spatial-social feedback revealed the interaction between Battery Park City's citadel design and resident attitudes. The impact of the citadel was significant but quite different than previously believed. Critics following Friedmann and Wolff's identification of the "citadel and the ghetto" warned that such spaces could not be authentically urban and instead would rely on a thin simulation of real city life. Richard Sennett concluded that "within this model community [Battery Park City] . . . one has the sense, in James Salter's phrase, 'of an illusion of life' rather than life itself."[1] Francis Russell likewise blamed Battery Park City's exclusiveness for filtering out "the 'exhilarating' aspects of urban life," so that "existence in Battery Park City will be very pleasant, but it will never constitute the same kind of rich urban milieu that makes Manhattan what it is."[2] To the contrary, residents' deep involvement in political contests over redevelopment suggests that concern over debilitating effects of the citadel on residents was misplaced. My observations found residents to be very actively engaged in community affairs; the danger was that residents had come to constitute a formidable new bloc that defended the citadel's exclusivity against challenges to make the neighborhood more accessible. Rather than impair residents' ability to come together as a community, the citadel's barriers shaped residents' views on the redevelopment projects just as it shaped the definition of the community itself.

The battles over West Street and the parking garage were unique opportunities to assess long-standing criticisms of Battery Park City in particular and citadels in general, that they were exclusive, elitist places, Disneyfied simulations of real urban spaces that filtered out the diversity and grit of the city. Once again, exclusivity was both a preexisting condition and a relationship of inequality that residents sought to reproduce. Despite concerns about whether neighborhoods like Battery Park City are "real," Battery Park City is a very real representation of what I have called the suburban strategy. As global citadels today take a page from the suburbs and maintain inequality through the establishment of great distance between the affluent and the less affluent, the citadel's walls become symbolic (though no less valued by residents for their symbolism). Inside those symbolic boundaries, the presence of anyone less affluent than these managers of global capitalism has been rendered unlikely. Battery Park City is, as critics have long asserted, a symbol of New York's shortcomings, but the danger it represents is not the boring simulation of Disneyfication but elites' further insulation through spatial isolation.

Thus resident views about these aspects of the redevelopment plan both further illustrate the reciprocal relationship of space and social relations and illuminate the problems—the use of suburban spatial strategies to defend residential inequality—inherent in the contemporary "citadel" first described by Friedmann and Wolff.

The Tunnel: Burying the Citadel's Walls

As guidelines for the World Trade Center site master plan were formulated, planners and New York City residents embraced the idea of burying West Street in a tunnel. Battery Park City residents' widespread and vehement opposition to the tunnel came as a surprise to planners, other New Yorkers, and city officials, who thought the plan would be welcomed for making it easier to cross the notoriously intimidating West Street. Planners outside Battery Park City concluded that West Street—or Route 9a, as state planners insisted on calling it (a name that accurately conveyed its being more a highway than a city street, but was foreign to city residents)—was "an eyesore, a failure of design."

After September 11, urban planners, embracing Jane Jacobs's emphasis on pedestrian-friendly cities, believed submerging the highway would make it easier for everyone to reach amenities and services on the other side of the street, improving quality of life for those on both sides and rectifying the "superblock" isolation of the old World Trade Center. A range of interested parties quickly saw advantages in burying West Street: traffic specialists had long been concerned about the dangers to pedestrians crossing a street so wide that few made it across before the light changed. The public approved of banishing mundane activities like work and driving from the Trade Center site so that the area could be completely dedicated to such high-minded purposes as memorializing, presenting high culture, and walking contemplatively down tree-lined promenades. The urban "growth machine" saw the opportunity to make a grand gesture as part of an unexpected redevelopment project. Downtown, decision makers like the Battery Park City Authority and the Lower Manhattan Development Corporation (LMDC) saw an opportunity to better connect the large business and retail cores of the World Financial Center and the rebuilt World Trade Center. The tunnel idea gained popularity, and when the LMDC issued guidelines for the Trade Center master plan competition eventually won by Studio Daniel Libeskind, it required all applicants to cover over West Street. The issue became the purview of the state's Department of Transportation (DOT), which drew up four designs for West Street. The DOT was much more oriented to traditional highway planning statewide than to the needs of the country's most pedestrian city. They emphasized that a tunnel would bypass intersections and speed drivers on their way, reducing emissions from idling traffic. The initial plan, called the "long tunnel," would have kept cars underground for the whole length of Battery Park City. The DOT soon abandoned that plan in favor of their second proposal, a "short tunnel" that would cover traffic only along the length of the Trade Center.[3] The third and fourth plans, often considered mere foils, left West Street at grade level. It was in response to these proposals that residents organized.

Residents first learned of proposals to bury West Street during "Listening to the City," a massive meeting of New Yorkers to propose and discuss ideas for the rebuilding of Lower Manhattan after September 11. Some attended roundtable discussions held in July 2002 at the convention center, while others participated online via a Web page set up to prompt discussions and solicit people's opinions. At these meetings, people were asked to respond to several ideas, including a proposal to replace West Street with a tunnel. Once the street was covered, designers variously proposed using the "new" space for a park, a memorial walk with a view of the Statue of Liberty, or new office buildings.

Battery Park City residents immediately and broadly opposed the tunnel plans. In an interview at the white-shoe law firm of Dewey Ballantine,

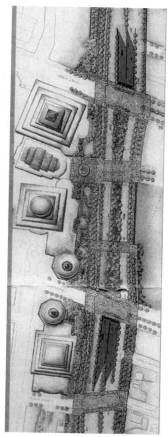

Left: West Street, also known as Route 9A, and as a "failure of design" (New York State Department of Transportation brochure). *Right*: The state Department of Transportation's plan for a "short tunnel." Triangular shadows indicate the entrance and exit ramps on West Street. The geometric shapes to the left are the tops of the World Financial Center buildings. The tunnel itself would have stretched only the length of the World Trade Center site. Plans called for more crosswalks than had been there previously, because the tunnel would have been covered by a new street. (New York State Department of Transportation, "Route 9A Lower Manhattan Redevelopment: Project Information," 2003; author's private collection)

where he was an attorney, Battery Park City resident Bill Love explained: "In 'Listening to the City,' outside residents said, 'Yeah, bury it.' That's when I knew we had a problem." Tom Goodkind, a financial officer at a real estate firm who was a Battery Park City resident and Community Board member, later argued to state traffic planners during one of many public presentations of plans for West Street that "basically this is an amenity for the tourists and for traffic . . . but I would like to hear what this does for *residents*. And I've asked you every time to come and address residential issues. I'm not sure if you've been listening to us at these—what we call dog and pony shows." Any resident who spoke publicly about the tunnel opposed it. As one Community Board member recalled, nearly forty people had spoken against the tunnel at a single meeting. A rare nonresident member of a Battery Park City community group observed that "the one thing that unites them, more than the tunnel, is that they feel intruded upon by the outside world. It's kind of like them against the world. It's all these outsiders trying to build their community for them. I don't really disagree with [residents' concerns], but they're so blinded by that, that if the outside world said let's rebuild West Street the way it was, they'd say, 'No, let's build a tunnel.'" While the speaker may have exaggerated residents' position, he captured how residents felt threatened and besieged by outside plans, even those that the planners claimed were for the benefit of the privileged community.

Residents discussed the tunnel plan on an online message board that had been heavily used since September 11, first to find fellow residents and subsequently to address neighborhood problems. After an exchange of messages about the tunnel plan, Joanne Taylor proposed meeting in Battery Park City's Rector Park to "consolidate and organize our opposition to the proposed West Street tunnel." Tired of mere input, the group sought actual influence.

Though their discussion had begun on an online bulletin board, the one founding meeting of the Coalition to Save West Street in Rector Park cemented the group geographically: in my interviews over the course of the following year, I found that most of the core members lived near Rector Park. Of the six most active members, four (including a couple) lived in the buildings that actually faced tiny Rector Park. One other lived in Gateway, and a sixth, exceptional member lived in Greenwich Village. (He later became less involved.) After the meeting, Taylor commented that "the meeting was well attended by Rector Place residents, but where were the rest of you?" That concentration remained a feature of the group, even though most of their subsequent communication and mobilizing was done online, and even though, by every measure taken afterwards, opposition to the West Street tunnel was dramatic and widespread throughout Battery Park City. Like the groups discussed in the previous chapter, space played a formative role for the coalition. While other studies of community organizing have illustrated its local nature, it is significant that online communication and the supposed despatialization of politics inherent in the Internet did little to disperse this group.

Not all residents, of course, opposed the tunnel, but the momentum of neighborhood activism was almost universally against it. One resident activist believed that "98 percent of people who lived here before [September 11, 2001] are against it." Surveys found strong opposition to the tunnel.[4] In the most discussed poll, from May 2003, 67 percent of Battery Park City residents expressed a "negative" opinion toward the tunnel. This was higher than other Lower Manhattan neighborhoods' responses.[5] Two other polls found comparable percentages of Battery Park City residents opposing the tunnel.[6] Not only were most residents opposed to the tunnel, but no community activists advocated for the tunnel that much of the rest of the city, and the redevelopment agencies, believed was in residents' best interests.

Residents' opposition to the tunnel illustrated the influence of Battery Park City's exclusivity on their opinions about local development plans. Depending on the setting, residents articulated one of two different sets of objections to the tunnel. When speaking to a citywide audience or as representatives of a community group, they worried that the tunnel would have the opposite of its intended effect: the tunnel would actually cut off Battery Park City. Noise, dirt, construction, and traffic from the tunnel would harm the neighborhood. The cost, at least $800 million, was too high.

In discussions within Battery Park City, however, the same residents could prioritize different, even contradictory objections based more firmly in exclusivity, worrying that covering up West Street would eliminate a barrier that isolated them from the rest of the city. Residents wanted to maintain the citadel's exclusivity. On one of the online bulletin boards for Battery Park City residents, participants discussed the tunnel plans for several weeks.[7] Here residents embraced Battery Park City's "suburban feel" and claimed that "West Street provides a barrier from the rest of the city, and will help us to retain some of the character of our neighborhood." Another admitted, "I do not want West Street buried because I like the feeling of being separated from other parts of the city. This is, I know, a totally selfish reason." Thus, while tunnel advocates outside Battery Park City painted West Street as a nuisance that the tunnel could correct, in public residents described the tunnel itself as the nuisance. In private they reversed their assessment once again, embracing West Street precisely because it had the obstructing effect tunnel advocates sought to alleviate. In this view the tunnel, once completed, would not be enough of a nuisance.

The Citadel's Effects on Residents' Attitudes

That residents immediately opposed the tunnel plan derives directly from the role West Street played in demarcating their community and shows the effects of the citadel form on residents' attitudes and views toward their community and the city at large. To residents, the difficulty nonresidents in particular experienced crossing West Street was a virtue to be preserved, even at the expense of greater promised convenience for residents. Not only did the barrier function of West Street heighten the neighborhood's sense of seclusion, it had become a fundamental element of what distinguished their community in the eyes of residents. Residents' actions in response to the tunnel plan also show that beyond corporate and government elites already identified in citadel critiques as sources of urban inequality, citadel residents are the most active defenders of their community's privilege.

The public reasons residents gave were not simply pretexts. Concerns about construction or the disruption of their everyday lives were real: proposals to accelerate the construction schedule (and reduce traffic disruption) would have required increasing the number of days and hours per week when noisy construction work was under way. On the other hand, shortening the workday would extend the number of years during which getting in and out of Battery Park City would be difficult. Yet the symbolic role West Street played in keeping anxieties about the city at bay was more important than the practical inconveniences: that symbolic role linked the type of life residents enjoyed in their neighborhood to the preservation of the West Street highway barrier. Battery Park City residents old and new opposed the tunnel in much the same way, overshadowing other perspectives held by people in work or volunteer organizations.

The strength of Battery Park City's influence on residents' views is typified in the opinions of one new resident whose company had relocated him with his family from the suburbs to Battery Park City soon after September 11, 2001, to oversee the firm's return to the World Financial Center. His family adapted quickly to urban life; I had met him when both our children were at the playground. Less than a year after the attacks, he explained his opposition to the tunnel plan just as many others in the community did: beginning with objections to construction ("The noise and disruption, why do you want to make things harder down here?"), then quickly shifting to a discussion of the valuable barrier function West Street played in secluding Battery Park City. He believed there was something special about Battery Park City that he wanted to preserve. While admitting that his opposition to the tunnel made other people think residents of Battery Park City were snobs, he still believed that "there's a reason to protect it from the rest of the city." After only a few months in Battery Park City, this resident's views diverged from the widely held view among outsiders that the tunnel would be beneficial to the community, and mirrored the locals' belief that the West Street barrier was an asset, not a liability. As this resident demonstrated, people adopted a defensive posture toward the neighborhood's isolation quite soon after moving in.

Resident Joanne Chernow was so welcoming to tourists generally that she volunteered in a program that greeted visitors to the city. Yet she framed much of her opposition to the tunnel in terms of the negative effects of tourists downtown, including the fear that they would soon come to Battery Park City in large numbers. For Chernow, tourists and memorials were emotionally upsetting. Like other residents in the first year after the attacks, she said she wanted to get back to "normal" and found the presence of visitors and memorial shrines disruptive. She described visiting the site of the disaster as "ghoulish," saying that the presence of tourists was "just really preventing any sort of normalcy from returning. I think the bottom line is we want a normal neighborhood, we want a livable neighborhood. There has to be a focus on something positive." To her, West Street had always played a barricade role: she volunteered that she and her husband used to joke that if West Street "wasn't there, we'd have a lot more scum over here." (This resident would not specify who was "scum," though she followed up by saying that she had heard homeless people had spent the night in the Winter Garden Mall, so she didn't know how effective West Street actually was.)[8] While residents had particular reasons after the trauma of September 11 to be reluctant to have large numbers of visitors in their neighborhood, their attitudes towards those visitors were formed before September 11, in a community that normalized neighborhood exclusivity in the midst of a much more publicly accessible city.

Residents were not ignorant of how their opposition to the tunnel appeared to outsiders. Though he was critical of many aspects of Battery Park City, Manohar Kanuri expressed his opposition like most others, first arguing that the tunnel was not necessary because Battery Park City was already connected to the rest of Manhattan, then defending the status quo for the way it isolated Battery Park City. Involved in finance and real estate before and after September 11, in the in-between years he spent more time on activist and political pursuits. Kanuri sought to link his community activism to progressive politics that opposed the wars and jingoism that followed the attacks. In his home in rent-regulated Gateway Plaza, perched between the Trade Center's Ground Zero and the luxury yachts of the North Marina, Kanuri explained,

> We're kind of not really in favor of [burying West Street] simply because, what is the major reason justifying it? To reconnect Battery Park City to the rest of Manhattan. Hello? We have been living here for ten years, we've been connected to the rest Manhattan. There have been no deaths of people crossing West Street. If you feel more connections are needed, sure, you can put more connections. Burying West Street will cost billions, will take years, there will be overruns, quality of life will be zero.
>
> We had the best of both worlds, really. . . . So people living in Battery Park City had that sense of community simply because there was a barrier dividing Battery Park City from the rest of Lower Manhattan. What was the nature of that barrier? It was a simple highway. And there was no reason for traffic to go through Battery Park City. If you look at any other neighborhood in Manhattan, you're talking about traffic going through the neighborhood.

Battery Park City's reliance on the barrier of West Street as a key source of its distinctive quality led residents to oppose breaches to the barrier. The tunnel plan also brought out a recurrent theme in conversations with residents: regret that their "secret" was becoming better known and that more people were using the existing park. Particularly given the financial contribution of the rest of the city to what one observer pointed out "is, after all, a publicly sponsored luxury housing development," their dismay seems out of place. Nonetheless a large source of the opposition to putting a park over West Street was that parks are attractive to people. Kanuri warned:

> If it's all parkland, you'll have half of Manhattan right there. Just over the last five or six years, the promenade has become very popular with people from all over Manhattan. Ever since the *New York Times* started writing about it, more people started coming by on weekends. It's not the same

thing. If you want to take your kids and walk around the promenade, which is why a lot of people wanted to come here in the first place . . . it's very cramped and crowded, you have rollerbladers streaking past, you have cyclists ringing their bells, going by at fifty miles an hour. I think that gives us an idea of what's in store if West Street is turned into either a park or office buildings, because either way it changes the relation to the barrier.

But Kanuri, like many other residents, was familiar with the history and reputation of community opposition, and he observed that "the problem with articulating these things, at least for me, is, [people say,] 'You're just being elitist, not-in-my-backyard idiots.'" He was aware both of Battery Park City's reputation and how the community's position was likely to be interpreted outside the neighborhood. This recognition led him to articulate an alternative perspective on that perceived NIMBYism: "I would say to these planners, instead of trying to integrate us to the rest of Manhattan, try to replicate us in the rest of Manhattan, to tap into this sense of community." But while Kanuri denied, with this recommendation, that he sought to deprive others of what he sought for himself, he remained true to the vision the citadel proposed of community defined by its barrier.

Kanuri at once captures and complicates residents' position. Like others, his objections move seamlessly from the public concern (over construction noise) to the private preference for seclusion. But he is aware of the risk of being dismissed as a NIMBY elitist and provides an alternate explanation. As he suggests, there are positive attributes of Battery Park City that are unlikely to be extended to other neighborhoods, and he invites the listener to consider Battery Park City's strengths rather than its stereotypical exclusivity. Views from the citadel are not one-dimensional. Kanuri demands that outsiders consider more seriously both the advantages and the voices of Battery Park City.

Active members of the community denied that their opposition was based on any desire to keep the neighborhood isolated. John Dellaportas (like Bill Love, an attorney and member of Save West Street) cautiously acknowledged the presence of exclusionary sentiments. People he had spoken to through Save West Street had one of two responses to the proposed tunnel: either "it's not going to reconnect us, or two, though this is not our position, we don't want to be reconnected." Ben Hemric, another member of Save West Street, explained that advocating for the continued isolation of Battery Park City by West Street was not a position that could be taken publicly. "That's sort of a taboo topic. I think about three-quarters of people who live in Battery Park City feel that way, but only one-quarter or one-eighth . . . [would say so]. Even the chairperson of the Community Board has stigmatized that position as being elitist, or—I'm looking for the right word—as a *suburban mentality*, because they have their gated community and don't want other people to intrude."

In other cases, respondents denied that West Street served as a barrier by considering only whether it was a barrier to residents like themselves, not to outsiders. They did not consider that those who called the street a barrier might actually be complaining that it kept out nonresidents. An account by Bill Love of his own interactions with West Street was telling. Love stated that before September 11 the street "was not a barrier to crossing." But before moving to Battery Park City eight and a half years earlier, he had visited the World Trade Center several times and, after getting lost in the underground mall, would end up, without knowing it, facing West Street. "A couple of times in my wandering, I'd come to One World Trade Center, and I'd be at this really busy street, and there were cars whizzing by, and I'd see that neighborhood and just turn back. *But for someone who knows the neighborhood*, it's easy to get over there." He had experienced Battery Park City as impossible to access when he had been a nonresident, but once he lived there himself he no longer acknowledged the difficulty of public access as a problem.

Longtime resident and activist Linda Belfer's denial that West Street was a barrier is a striking example of how physical space comes to be seen by people as natural and inevitable. "Well, personally, I don't see it's a barrier. It's *just an eight-lane highway* that we had to manage to get across one way or another." This led her to mention Pearl Scher, a fellow Battery Park City resident and member of Community Board 1 who was still suffering from being hit by a car on West Street, but she did not revise her assessment of the street. More striking still was the incongruity of Belfer's views with the fact that she used a wheelchair and had been frustrated in her attempts to attend several Community Board meetings because the elevators at the pedestrian bridges over West Street didn't work. Some bridges she had never been able to cross. Yet the disconnect between people's experiences and their concept of the space is not at all uncommon. People accept physical space as it is and don't consider it to be malleable in the way other manifestations of social relations are. Belfer, for instance, had spent more than twenty years strengthening the neighborhood's political position in Community Board 1 and changing her fellow tenants' relationship to the landlord through her work in the tenants' association. Regarding landlord-tenant relations she didn't assume things could not change, but the physical environment appeared unmalleable.

Kanuri likewise denied his own experiences of West Street as a barrier and expressed the common suspicion that the promise to reconnect Battery Park City was just political pretense: "It wasn't a problem until somebody made it a problem. And I want to know who made it a problem. Battery Park City has been around for twenty years. I'm sure some people didn't find it too convenient. My mother for instance, didn't find it was easy to cross West Street by herself. But . . . there's always something that bothers you about a city." Although Kanuri's own mother had trouble crossing West Street, that didn't make it a barrier, nor did he think there was much to be done about the inconvenience.

In these ways, residents' objections to the tunnel were inextricably tied to their definition of the community through the exclusionary function of West Street. That the neighborhood's exclusivity was maintained spatially gave it a naturalness that segregation based on socially maintained rules, like private guards and signs reading "Residents Only: No Trespassing," would not have had. While the tunnel plan was seriously flawed for other reasons, Battery Park City's strength and distinction would easily have weathered a better connection to the rest of Lower Manhattan. But the central importance of West Street's inaccessibility in residents' conception of the community ensured that community action would be uniformly in opposition to a tunnel.

No community can be unanimous, but opposition to the tunnel was hegemonic enough among community members that those who were even willing to consider a tunnel expressed that view only in private conversations. Back home in the apartment she had fled on September 11, one veteran of previous community battles could imagine a version of the tunnel she would support, but she expected that greedy developers would force an unacceptable design that gave them room for more large buildings. She confided, "There's a whole group of people who are adamantly opposed to the burying West Street from many different perspectives—not the least of which is the cost factor, and they feel as if it would isolate Battery Park City . . . I fear that any land that would be created as a result of that . . . that greed would indicate big tall buildings should be built on it. If it would remain public open space, active or passive recreation space, and we had that assurance in perpetuity, and we had assurance that they would make sure that Battery Park City would not be isolated, you know, then I could live with it. But many people don't agree with me." Another publicly active resident kept such conditional acceptance of the tunnel private as well, hedging its expression to me with cautions: "My personal preference, and this is a matter of personal preference, it may not even be realistic. I'd love to see it. I'm hesitant to express my opinion because the objections to a tunnel are very real . . . [But I'd love to see] a tunnel below, and a parklike boulevard above, and retail space, maybe a library." These residents never publicly shared their cautiously optimistic visions of what a tunnel might mean. Both were active in the community and were proud of being independent thinkers. But they were also well integrated into the networks of the community's politically active residents and didn't want to upset their peers by making public statements that would draw the ire of those who were actively mobilizing opposition to the tunnel plan.

Thus citadel residency and the physical isolation of Battery Park City outweighed other considerations, solidified public opposition to the tunnel, and led residents to articulate that opposition in terms of the value of Battery Park City's exclusivity. This influence of the citadel on residents and its reinforcement of exclusivity point to a significant way the social exclusivity of physically exclusive neighborhoods becomes self-reproducing.

Despite the community's relative homogeneity, Battery Park City activists were forced to address conflicts and disagreements between local organizations. Tunnel opponents did face conflict with groups who were not oriented wholly toward Battery Park City. Support for the tunnel from Community Board chair Madelyn Wils (a resident of neighboring Tribeca) provoked resentment and quiet calls for her ouster. Residents in Battery Park City and Tribeca shared many similarities: income and racial distribution, residence in Lower Manhattan, experiences of the collapse of the Twin Towers just next door, and the difficulties of living in the "frozen zone" in the months that followed. They even shared an identity as pioneers who had come down to an area where no one lived (Battery Park City was empty at about the same time Tribeca's light industrial warehouses had few residences). But Battery Park City residents quickly ascribed Wils's initial support for a tunnel to the fact that Wils lived on the other side of West Street. Occasional residents even grumbled that the local Community Board was like an "occupying army" and suggested that Battery Park City needed to organize separately from other parts of Lower Manhattan.

Wils's initial support for the tunnel was taken by members of the coalition as evidence that Community Board 1 couldn't represent Battery Park City, even though at the same time one of the same members said that Anthony Notaro, chair of Community Board 1's Battery Park City Subcommittee, and a Battery Park City resident himself, was "open to the tunnel." But while views of those two board members did not vary dramatically, Wils's views were construed as evidence that outsiders were different, while Notaro's were overlooked.

The WTC Residents Coalition (which was founded in Battery Park City but explicitly reached beyond its borders) avoided taking a formal position regarding the tunnel because members outside Battery Park City supported the tunnel. Such prevarication was unacceptable, and one Battery Park City resident complained that the group was "wishy-washy" about the tunnel. Eventually groups were able to negotiate with the WTC Residents Coalition to join the opposition. In such disagreements, community groups experienced some of the conflicts all neighborhood groups do when trying to build support for their position. Far from living an insulated, simulated life, residents were arguing and negotiating real political differences.

Similarly, activists worked to gain support from local politicians and other community groups. The local state assemblywoman, city council member, and congressperson all came out in opposition to the tunnel in the next year. Save West Street then sought endorsements from other community groups; in the following year both Battery Park City's largest tenants' association and a post–September 11 residents' group allied themselves with tunnel opponents.

Activists used events and public spaces to publicize their opposition to the tunnel. In 2003, I walked through the Second Annual Battery Park City Block Party. Thousands of people milled about on a street wedged next to the office towers of the World Financial Center. Tables were set up for people to eat food from the local restaurants. A resident urged me to sample a local company's spicy pickles. Many attendees were Battery Park City residents. The crowd tended to be upper middle class in appearance, predominantly but not entirely white. Some adults had children, but plenty did not. A large mobile stage bedecked with the insignia of corporate sponsors and the block party's new logo was installed, where local residents performed songs, children's groups danced, and music played over the sound system. A salon gave outdoor haircuts and a painter set up her easel. The event proved useful for local groups: Save West Street and others handed out flyers from their tables and talked to residents. At the end of the afternoon, Rosalie Joseph brought several of the children's dance troupes back on stage for a closing song. At the end, the children in front held up a banner that read, "You Can't Stop BPC!"

Top: The Second Annual Battery Park City Block Party, September 2003. Performers make explicit the block party's community-strengthening mission. (Photo © Robert Simko.) *Middle*: As seen from the stage, a much larger crowd than most block parties attract. *Bottom*: Organizer Rosalie Joseph addresses the audience.

The sign captured organizers' and residents' message that the community would come back and would be strong in the face of all challenges. The message's optimism was welcomed enthusiastically by residents, and two years into a very uncertain recovery process there was no apparent over-reaching in positioning Wall Street's bedroom community as a spunky underdog. For the block party, as in opposing the West Street tunnel, residents had done the traditional community organizing work of building a coalition, gaining supporters from politicians and other groups, and promoting their position and soliciting supporters at events in Battery Park City's public spaces. As with any other community, residents' efforts bore fruit in a strong sense of community, but in Battery Park City this was further reinforced by the power inherent in being such a strategically positioned community.

That year the full body of Community Board 1 passed a resolution opposing the construction of the West Street tunnel. At a meeting that summer, the Community Board listened to statements by sixteen speakers representing developers, local politicians, residents, and resident organizations. All opposed the tunnel plan. The Community Board (which covered Battery Park City and several other neighborhoods), voted 25–10 opposing it.[9] (The board's chair, who had previously supported the tunnel and was a member of the LMDC board of directors, abstained.) That long-sought-after resolution, a culmination of persistent lobbying and the collection of endorsements by a range of local actors, was celebrated as a victory by community groups. But the state's tunnel plans remained unchanged.

Little was said about the irony that the Community Board opposed one tunnel only to endorse another: In the same meeting that voted against the West Street tunnel, the board proposed applying the nearly $1 billion budgeted for it to a $6 billion project to directly link Lower Manhattan to Kennedy Airport by train. The rail link would include a three-mile-long tunnel under other parts of Lower Manhattan and Brooklyn neighborhoods. The barrier of West Street was one step closer to being preserved, and the way that barrier restricted activists' areas of concern—to the tunnel next to their homes, not the proposed tunnel a few blocks away—was that much more evident.

Though residents followed the bureaucratic progress of the tunnel in detail—attending environmental impact statement meetings, dividing and reading thousand-page reports, and speaking at Community Board meetings, activists saw the ultimate decision as resting in the hands of the governor, to whom key state actors like the DOT and the LMDC answered.

After almost two years of intensively lobbying politicians, voicing opposition, and getting residents to speak at hearings, an active member of the Coalition to Save West Street worried, "Out entire fate still lies with the governor. Doesn't matter how many resolutions, or groups. If the governor says he wants a tunnel there's going to be a tunnel." As late as December 2004, a year and a half after the Community Board resolution, it was reported that the governor still supported the tunnel plan and was even considering making it longer so that its access ramps would not surface in front of a planned Battery Park City skyscraper for the investment bank Goldman Sachs. Four months later, Goldman Sachs suspended negotiations over the property they were expected to lease, citing the tunnel plan, which the firm believed would interfere with the office building. The next week, the governor abruptly rejected plans for a West Street tunnel, and the DOT endorsed an alternate plan to relandscape and reconfigure West Street at surface level. The battle for West Street was suddenly over, though there was no public evidence that community activism, as committed as it had been, had made any difference.

Even though Goldman's opposition precipitated cancellation of the tunnel plan, residents had set the stage to make such a decision easy for the governor. Pressure against the tunnel plan had been growing, and its justification seemed dubious. From the perspective of interest group politics, few people were invested in actively promoting the tunnel. Planners inherently liked the idea, but they were not an organized, lobbying force. The only financial beneficiary would have been Brookfield Properties, who would have benefited from having their World Financial Center better connected to potential customers. But other nearby building owners gained nothing from the mammoth plan. In opposition were well-organized residents, with every politician signed up on their behalf. The boards of the LMDC and the Port Authority were heavy with gubernatorial and mayoral appointees, making it more likely that they would support whatever decision the governor made than that they would lobby intensively for one position or another. Had the plan been well under way with broad political support from influential players in the city, Goldman might have been able to alter, but not cancel, the plan, or might have used the state to arrange favorable terms on another piece of Downtown property. But residents had stripped away tunnel support, leaving no compelling reason for the governor to insist upon it in the face of Goldman's concerns.

Whatever the role of community mobilization in altering the tunnel plans, activism was much more substantial than critiques of the anemic life of citadels, Battery Park City in particular, would have predicted. After all, it is not uncommon for community opposition groups to lose their struggles.[10] But the role of Goldman Sachs illustrates one significant difference for citadel communities: the citadel was home not only to residents (who themselves enjoyed greater political access and agency than most people) but also to even larger, institutional actors that dwarfed residents' influence: Battery Park City contained, among others, the New York Mercantile Exchange, large offices of American Express, Merrill Lynch, and Deloitte and Touche; it would soon include the headquarters of Goldman Sachs; and the neighborhood was run by the governor's appointed representative. Thus citadel residents were more likely to see their goals abut objectives of more powerful players. The collocation of corporations meant that resident power would be either greatly magnified or negated by neighboring corporate power. In this case plans that were allegedly intended to make Battery Park City less exclusive were defeated by corporate power. Though sharing the neighborhood with corporations might in other situations stymie residents' efforts to shape their community as they saw fit, wealthy citadel residents were more likely to be allied with local citadel corporations' agendas than opposed to them. In such cases residents enjoyed a powerful multiplier effect by aligning their goals with those of their business neighbors.

The Parking Garage

By the time the tunnel had been scrapped, another component of the redevelopment plan was gaining Battery Park City residents' attention. Plans for a bus garage illustrated more ways in which the spatial and exclusive definition of the community shaped residents' views.

The centerpiece of the March 2003 meeting of Community Board 1's subcommittee on World Trade Center Redevelopment was a presentation by the Port Authority on the state of plans for the World Trade Center site. Because of the level of public interest, the meeting was held in the hearing rooms of the powerful State Assembly speaker Sheldon Silver. Because his district and offices were located minutes from Battery Park City, the rooms provided both an august setting for a community board meeting and an easy way for local community groups to stay connected to the assemblyman when meetings were expected to draw overflow crowds. I was in the audience with about a hundred residents who attentively followed a PowerPoint presentation that covered everything from an extensive underground transit terminal that would connect a half-dozen subway lines and commuter trains, to new ferry terminals, to grand public spaces around the site, to the complex nest of service components that would be located below ground. But as the floor was opened for questions, first from Community Board members and then from the public, the impassioned discussion was about only one thing: plans to park buses on the footprints of the Twin Towers.

Building anything on the footprints was bold. If the towers themselves were not to be rebuilt, as some people had proposed, then in most plans and discussions that space was imagined only as a memorial. Indeed, the winning plan by Daniel Libeskind proposed an open memorial on the footprints of the two towers, extending down to the concrete foundation built on bedrock that was exposed seventy feet below street level. But the Port Authority had already quietly moved ahead with reconstruction of the PATH train tracks where they had originally been beneath the South Tower.[11] Representatives of several groups of victims' families had reluctantly gone along with that plan, but using the space as a parking garage felt to them like a sacrilege.

Contrary to the wishes of family members, Battery Park City residents at the meeting were firm: they insisted that the parking garage *must* be built on the footprints of the World Trade Center towers. "There is no other place," said Community Board 1 member and Battery Park City resident Jeff Galloway. Later, he added that what mattered was not what people thought now but what the memorial would make future generations think seventy-five years from now—and they owed it to future generations to put bus parking for the memorial site on the footprints. One after another, residents insisted that the parking garage be built under the memorial.

As with plans to bury West Street, residents had reasonable concerns about the parking garage. They feared that the 160 buses per day projected to come to the memorial would overwhelm the neighborhood. Already parts of Lower Manhattan, like other neighborhoods popular with tourists, suffered the loud noise and diesel exhaust of tour buses that idled for hours. The situation was at its worst during the summer when the number of buses peaked: residents complained that drivers squeezed their buses into any space they could find, whether parking was legal or not. Indeed in the summer it was not hard to find an idling bus whose driver kept the engine running to stay air conditioned while passengers were touring a site. Residents' open windows left them exposed to the noise, dirt, and hot exhaust of the buses.

A parking garage for the site was important. What was never clear was why it had to be on the footprints themselves. Of the sixteen acres of the World Trade Center site, only about four were part of the tower memorial (the tower footprints themselves were about 2.5 acres). Since nothing had yet been built above ground on the site, there was other space on the site where a bus garage could be built, either on its own or as part of an office building. (Office buildings often have parking structures underground or on above-ground floors.) But residents never asked why a parking garage couldn't go elsewhere. They insisted—vehemently, and against family members' objections—that it be built under the memorial.

The opposition between the groups called "residents" and "families," and residents' sense that they were ignored in favor of the families, informed the discussion at the Community Board meeting from the start. The meeting presented a dynamic completely at odds with public perceptions that in the somber and respectful months and years after September 11 victims' families were accorded boundless consideration and their preferences were sacrosanct. In the actual meetings at which designs were debated, family members were openly assailed. "It's time to consider the thirty thousand people living in Lower Manhattan," said Diane Lapsum of Independence Plaza, to loud applause. Before long, the mood in the room turned hostile, and speakers directed their comments not at representatives of the Port Authority, but at the two or three "family" members sitting together. One attendee warned, "To the families: You can't put together the pieces of your heart by cutting out the heart of our community." There was little actual conflict, since the family members were so thoroughly outnumbered. Residents did most of the talking. Addressing families' fear that the bus garage might make it easy for terrorists to get a bomb under the memorial, one person goaded, "If security is your only concern, it may be another city in which one may want to live." She was, of course, speaking to people who had already lost family members to terrorism. Residents themselves experienced the anxiety of anticipating another attack like the one they had witnessed next door to their homes, but they belittled the concerns of others who had similar fears. As the tone became more heated, Helene Seeman, a member of Battery Park City United, offered what started off sounding like a conciliatory note. "You are victims, and we are victims," she said, as Diane Horning, mother of a victim, answered, "I agree. I agree." Then Seeman continued: "But *we* are also *heroes* for moving back." Preserving the footprints, Seeman continued, "is a ridiculous notion. And I'm not being disrespectful." Bill Love of the Coalition to Save West Street followed Seeman's comments by insisting on the necessity of a place to park the buses and said the footprints were "the only practical solution." (He added, in contrast to his comments about West Street during the tunnel debates, that the amount of space monopolized by the memorial already concerned him because "we're somewhat isolated in southern Battery Park City.")

As Manohar Kanuri had explained in an interview, fellow residents found it was difficult for their priorities to compete with those of family members, who seemed untouchable because of the losses they had suffered. So Lower Manhattan residents drew on earlier tragedies they thought gave them standing to reprimand families for wanting nothing built on the space. One man at the meeting said his family had died in the Jewish ghettoes during World War II. "I'd like to go to Warsaw and knock down some of the blocks where my relatives were destroyed." For family members to want the World Trade Center site empty, he said, was selfish and unrealistic. Another man said he had "rolled up bodies in ponchos" as a Marine officer in Vietnam in 1968 as a prologue to his insistence that "there is no hallowed ground" and that any construction could take place on the site. "You don't own those sixteen acres," he told one of the family members near him.

After participating in many such meetings by that point, I understood why residents wanted a garage for the buses, but it was mystifying to me why they felt it needed to be on the footprints of the tower. Nor was I prepared for the vitriol directed toward family members. Residents had said in a variety of contexts that they felt the family members were given too much deference—by planners, by the media, and by the general public—and that residents' needs were ignored. But I still did not anticipate the anger displayed at the meeting, nor have I ever understood why residents directed their frustrations at residents, who had no say over the placement of the garage after all, rather than questioning the Port Authority or seeking assurances from those officials that the need for a garage would not be forgotten. The possibility of a breach in the neighborhood's exclusive boundaries could always incite some anxiety, and the building resentment toward families accounted for some of the reaction as well, but there may have been additional fuel. This was one of several occasions when public meetings about memorializing that horrible day suggested that emotions such as mourning, guilt, anger, frustration, and denial not only were individual, internal phenomena but could manifest themselves collectively and publicly, through a community or organization as well, often without participants fully recognizing the force of the emotions that had developed in the room.

No one sought to temper the tone of the meeting or the attacks against family members. No Battery Park City residents spoke out against the parking garage. Nor did any ask the Port Authority why it had put the garage, which residents felt was so important, in a politically contentious place that left many residents fearing it would not be built at all. Nor did anyone seriously pursue with the Port Authority representatives at the meeting whether alternative sites were feasible.

Six months later, the Port Authority itself proposed alternative sites. At the September 2003 meeting of Community Board 1's World Trade Center redevelopment subcommittee, during a question-and-answer session representatives from the Port Authority revealed that the Authority was now considering building bus parking either just south of the Trade Center site or further north at "site 26," a vacant lot on the Battery Park City side of West Street that was already used as a parking lot. The new plans would create parking garages for the buses without running into opposition from family members trying to protect space for the memorial.

The response of Battery Park City residents was just as surprising as that at the first meeting, underscoring the role of West Street in separating Battery Park City from the rest of the city. Had the depot been built on the footprints of the Trade Center towers, it would have been across the street from some downtown residences and only about 750 feet from the closest apartments in Battery Park City, Gateway Plaza (home to a third of Battery Park City residents). It also would have been across the street from *southern* Battery Park City, the section where 89 percent of the participants in the Coalition to Save West Street's survey lived. And it was in southern Battery Park City that the coalition's leaders, spokespeople, and core members —Bill Love, John Dellaportas, Joanne Chernow—lived as well. Site 26, by contrast, was on the far north side of the commercial World Financial Center. It was almost 1,200 feet, or more than four and a half blocks, from Gateway and southern Battery Park City. It faced no residential buildings on any side. There were some apartments in northern Battery Park City, but they housed far fewer residents than Gateway did. The alternate bus garage site was much further from most residents of Battery Park City than the original site.

Yet members of the Coalition to Save West Street, who at the Community Board meeting had demanded a bus depot in the footprints of the towers as necessary for the neighborhood, circulated flyers at the next Battery Park City Block Party calling the bus depot on site 26 "a disaster." "Tell our officials," the flyer urged, "No tour bus depot! No West Street tunnel!"[12]

The Coalition to Save West Street makes its case against a bus garage in Battery Park City. The inset map on the poster shows three things adjacent to the ball fields: the planned garage, the planned tunnel, and the actual existing highway. No mention is made of the air pollution generated by the third element. (Flier, fall 2003)

The coalition offered five objections to the tunnel. Two applied just as easily to the garage at the other site: that it was "a new terrorist target" and "a $200 million waste of 9/11 funds." Two more of the objections might also be applied to the earlier site but, more importantly, were exactly the problems residents had earlier wanted a garage to *correct*: "engine noise and clouds of diesel fumes" and "traffic gridlock." The fifth objection, "danger to pedestrians and children at ball fields," sounded reasonable, since the two baseball diamonds were across the street from the proposed garage site. But the ball fields faced far more traffic and exhaust from their location right next to a busy section of West Street that the state had initially proposed putting underground in a "long tunnel" to reduce traffic and exhaust—a plan that was opposed by groups including Save West Street. Unlike most examples of community opposition to a noxious, undesirable, or nonconforming use in their neighborhood, Battery Park City's opposition was not to the bus depot: they had demanded that it be built. Nor were they a voice of not-in-my-backyard opposition in the typical sense, since they demanded that the bus garage be built *closer* to their homes, not further.[13] The only way to understand residents' positions requires understanding the role that space, landscape, and the design of their neighborhood played in residents' definitions of themselves and their community.

To residents, site 26 was closer than the World Trade Center tower footprints because site 26 was on their side of West Street, within the boundaries of Battery Park City. Residents further objected to building the garage on site 26 because it would bring buses—and tourists—into Battery Park City, whereas locating the garage on the footprints sequestered tourists in the memorial district. That space might be for the tourists, but the rest of Downtown should not be. Placing the garage on the Trade Center site reinforced the barrier role of West Street, whereas straddling the street with memorial functions weakened that divide. In this case, social space, not linear distance, determined the real proximity of the garage.

By the next summer much had changed, but residents were still unsure about whether they were being heard. Site 26 in Battery Park City was no longer being considered for the garage—it would eventually become the Goldman Sachs offices instead, and the firm had begun lobbying the state to cancel plans for the West Street tunnel. The garage was now scheduled to be built below ground a block south of the Trade Center complex, to the satisfaction of residents and families alike. Surface-level renovations to West Street began, but residents followed them warily, suspicious that official plans to make the area a "promenade" between the World Trade Center and the Battery's view of the Statue of Liberty would sacrifice residents' needs to those of strolling tourists. Because the strong boundaries of the community had given it a private definition, residents remained concerned about potential public uses of the promenade and other park spaces, which seemed fundamentally incompatible with the meaning they ascribed to those same places.

DIMBY and the Suburban Strategy

Battery Park City residents' contrasting responses to the two proposed locations for a bus garage showed how, given that residents opposed a site
further away from them, opposition that might otherwise be attributed to
NIMBY inflexibility is better understood with an appreciation of how the
built environment affect people's views. Their attitude was not NIMBY
but what I came to describe as DIMBY—Definitely In My Backyard—as
residents demanded that two things almost always defined as nuisances
—a highway and a busy parking garage none of them would ever use—be
placed in their backyard.[14] Appreciating residents' positions requires considering the neighborhood's history of isolation and the distinctive way in
which the spatial configuration of Battery Park City and its barriers affected
views toward local development proposals. Being able to look for DIMBY or
NIMBY tendencies adds the role of space to studies of community politics
and thus offers a fuller view of the framing of neighborhood movements.

That Battery Park City residents traded NIMBY for DIMBY suggests a
class difference. The image evoked by NIMBY movements is of middle-
class activists opposing projects because of environmental hazards. In the
case of the West Street tunnel, however, residents expressed no concerns
about the environmental risks of living next to a highway like West Street
(although the links between highway proximity and asthma were well publicized). The impetus to mobilize was not a desire to protect health but a
desire to protect status. NIMBY movements can also be about protecting
status, but dismissing proximate environmental hazards and focusing instead on status maintenance is at odds with the risk-averse image of middle-
class homeowners and parents and suggests a degree of confidence, even
hubris, more representative of upper classes like those that predominate in
Battery Park City.

Similarly, greater confidence is required to believe nuisance facilities can
be not just opposed but harnessed for one's own interests. In Battery Park
City, residents sought to deploy nuisances strategically to maintain exclusivity. Of course, middle-class neighborhoods are seldom faced with state
efforts to spend a billion dollars on a half-mile of roadway next door. Both
the level of state attention and residents' determination to influence, not
merely resist, those efforts contribute to the class specificity of the DIMBY
approach.

DIMBY remains counterintuitive enough to suggest it is rare, but the
presence in one neighborhood of two grassroots mobilizations motivated

by DIMBY raises the distinct possibility that further attention to space in community studies will uncover more examples of both the distinctiveness of upper-income neighborhood organizing and the strategic use of space in such ventures.

The use of space in Battery Park City to achieve social goals casts Boyer's critique of the global city's "spatial restructuring" in a different light. The citadel does represent a new spatial strategy of control, but it goes beyond the image of the contraposed citadel and ghetto. For while West Street shaped residents' conception of their community, it was not the dividing line between the haves and the have-nots. In terms of the spatial restructuring of inequality, the salient feature was not the barricade (which in this case separated residents from other people very similar to themselves) but space. The importation into the global city of the suburban strategy, in which classes and castes are separated not as much by walls as by distance, is well represented by Battery Park City. As first practiced in the suburbs, class- and race-homogenous neighborhoods are planned and constructed over large swaths of land, under the hegemonic assumption (made explicit by some of those who oversaw Battery Park City) that land use should be organized strictly on the basis of real estate value—increasingly, in cities, with rich residents closer to downtown and poorer residents much further out. This is a considerable reversal of New York's well-appreciated "block-by-block" diversity. And while that diversity is often overstated (see Massey and Denton on New York's "hypersegregation"), city planners' unapologetic elimination of the neighborhood diversity they had claimed to embrace in a post–Jane Jacobs world marks a significant alteration in the values of U.S. urban planning.[15] The story of Battery Park City can contribute to earlier critiques of citadels and the barriers that enable exclusion: though design matters, not walls but distance and neighborhood homogeneity are being used to reorganize the global city.

The citadel is central to understanding the social and spatial organization of the global city, but on close examination Battery Park City played a different role than the image of the citadel predicts. The citadel did not impair the intensity of community political life of its elite residents. The citadel's boundaries were integral to residents' perceptions of the privileges they enjoyed, so those boundaries influenced community positions on local planning issues and the organization and definition of community groups, and the boundaries, in this case West Street, became themselves an important focus of mobilization and activism. In this period of major reconstruction in Lower Manhattan, residents had a nearly endless list of potential community issues to address: the memorial design, office tower development in the World Trade Center site, transportation issues, the conversion of Wall Street commercial real estate into luxury lofts, chronically overcrowded schools, a war abroad, antiterrorism measures close to home, and more. That defense of the citadel's boundaries attracted the efforts of the largest number of community groups speaks to the social importance residents bestowed on this physical element. In this mobilization, the effect the citadel had on its community was to recruit residents as a resolute new interest group committed to maintaining the neighborhood's exclusivity. This adds the actions of local-level actors to the study of elites' roles in the global city. West Street has remained symbolically important, but its role could change: as elite enclaves have expanded until they fused with their neighbors, in practical if not symbolic terms individual citadel walls have become less significant in maintaining the new regime of spatial inequality than the segregation of city residents into developments like Battery Park City with narrow demographic profiles. Residents' means of maintaining that segregation, and the physical boundaries they use to do so, may change. But the role of space and the efforts of residential communities in the reproduction of the segregated global city are likely to continue. While residents of citadels may have proven to be committed defenders of citadel segregation, building an integrated city requires challenging the assumptions that shape such spaces.

8

Conclusion

The Suburban Strategy

The objective historian realizes that the 20th century is in transition to a remarkable new technology and a formidable new environment, before we have even learned how to handle the old one. Who's afraid of the big, bad buildings? Everyone, because there are so many things about gigantism that we just don't know. The gamble of triumph or tragedy at this scale—and ultimately it is a gamble—demands an extraordinary payoff. The Trade Center towers could be the start of a new skyscraper age or the biggest tombstones in the world.

—Ada Louise Huxtable, 1966

HUXTABLE WAS EERILY prescient, and not only in the disquieting comparison of the towers to tombstones. The twentieth century was not only an important historical landmark, both globally and nationally, but also a transitional period. Taken more generally, the gigantism that Huxtable observed—the oversized ambition of global capitalism—was a tremendous and hubristic gamble. The promise made by twentieth-century capitalism of greater material luxury, abundance, and modern convenience did indeed need to provide a tremendous payoff in exchange for the human and material toll it took on the environment and on capitalism's workers, bystanders, and consumers. The Trade Center towers were not the start of a new skyscraper age. Record-setting skyscrapers have moved on from the United States. (Eight out of the ten tallest buildings are in Asia. At this writing, the tallest building in the world is the Burj Khalifa in Dubai, twice the height at which the Twin Towers stood.) The attack on the World Trade Center was unimaginable when it happened, but the contention embodied by the towers themselves was not foreign to New Yorkers. Even when the attacks were raw in everyone's mind, a woman from Queens at a town hall meeting

Parachutists attached to giant flags drop through the sky during the 2008 anniversary of September 11.

on rebuilding didn't hesitate to tell me that "to me, from here, those towers always looked like two middle fingers pointed in the air." She did not want to see them rebuilt, because they represented to her the conflict between the global citadel of Downtown New York and the rest of the city, where she lived and worked. Others at the meeting disliked what the towers stood for with equal vigor. To these observers, miles from Battery Park City, the promise of downtown gigantism was broken.

June 2008 marked the closing of the frontier in Battery Park City, as construction began on the final two residential towers on the last empty lots. (In November 2005, Goldman Sachs broke ground on the last office building, after the West Street tunnel plan that the firm opposed was canceled.)[1] As these towers rose, the meaning of Battery Park City shifted from the potential people imagined for it to the reality of what had actually been built.

Of the many types of social structures people develop to reproduce social relations, the built environment is the only one that is concrete in the literal, not metaphorical sense. Changing any social structure can be a massive undertaking of organizing, politics, and social movements, but changing physical structure is a truly Herculean task requiring heavy machinery, engineers, and more concrete.

As the neighborhood's physical shape was locked in place, the community's social form seemed to be set for a long time as well. But while large-scale real estate developers and other members of the growth machine will move on from Battery Park City, the project of constructing New York is ongoing, and the process of determining the social and physical relations of American cities is just as continuously under revision. Battery Park City provides several important lessons regarding the suburban strategy that can inform developments that follow.

At many levels, Battery Park City was a success. It was a rare large-scale, planned project that had created a rich social community and vibrant public spaces. In the same era, many trends in American urbanism were based on the assumption that a project like Battery Park City couldn't work. Housing projects were being demolished as a form that hopelessly inhibited sustaining community. Jane Jacobs–scale neighborhoods continued to be gentrified, renovated, and historically preserved because homeowners believed community values were inherent in their tight-knit blocks. Advocates of "new urbanism" sought to reshape community by redesigning its physical form to replicate older, small towns. In comparison to these examples, a state-planned megaproject that had usable and popular parks and playgrounds, well-attended community events, and safe, walkable streets was a rare accomplishment. The community quietly challenged the idea that providing tall buildings, new construction, or large-scale urban development was a disastrous way to construct a community.

It may be hard to exaggerate the success of Battery Park City on those terms. What is easily exaggerated, however, is how much such a fresh start was needed in the first place. Toward the conclusion of my project, I met a wealthy retiree who lived in a gated community on a small island far from New York. She inquired about my project and then asked me, politely, if Battery Park City wasn't a positive development. It sounded to her, from my description, as though the planners of Battery Park City had found a way to make a "good" community, one that she might like to visit, in contrast to the horror stories about New York she had gleaned from years of media coverage. In those tales, New York is violent, chaotic, conflict-riven. It is not a place where people can raise children, where kids can go to school, where community can be relied on or personal safety expected. From my description, she thought Battery Park City might actually be a part of New York in which decent people like her could live.

Unfortunately, my interlocutor was mistaken. She was not mistaken that Battery Park City could be an attractive place to live. But decades of anti-urban discourse in the United States had convinced her that the rest of the city was not.

Though today millions of New Yorkers start families, raise and educate children, socialize at block parties, and live in safety, elites were convinced that New York's existing urban forms—of comparatively mixed-income neighborhoods, barrier-free access by outsiders, accessible public spaces, and occasionally integrated commercial streets—needed to be replaced. A new strategy was required to build New York. Battery Park City has provided firsthand insight into the construction, characteristics, and short-comings of that strategy. The suburban strategy was designed to create economically and racially homogeneous neighborhoods for upper-income professionals. With tax breaks that allowed it to be more lushly apportioned than the rest of the city, the neighborhood provided a fertile space for the construction of rich, sustaining community bonds. Even when put to the test by a harrowing disaster in its midst, Battery Park City's community proved resilient and supportive to those who stayed. But the very physical structure that allowed such an enclave also fostered exclusive attitudes among residents as a foundation of their definition of that community. Such exclusivity informed residents' views toward neighbors across the street, local redevelopment projects, and the rest of the city. The vibrant exclusivity of the suburban strategy encapsulates the achievements and shortcomings of Battery Park City.

With the adoption of the suburban strategy, not only in Battery Park City but, at the same time, across New York and in cities throughout the United States, comes several important assumptions about how cities should be formed. Plans to subsidize luxury residential housing for the benefit of corporate employers (and to not subsidize and integrate middle- and lower-income housing on the same sites) were justified by the nearly unchallenged notion that a space should be used in the way that generates maximum revenue. For the complex needs of human societies, this is a suicidal simplification. After all, streets do not generate revenue, and even toll roads do not generate the revenue that luxury property does. But luxury property needs to be accessible by roads, or it will be worth nothing and will soon burn down when fire trucks can't reach it. Likewise, vital public services, such as fire, police, health care, and education, produce no revenue at all but are needed everywhere by everyone. Shunting working people, working-class jobs, public services, and public spaces to the financial and spatial margins because of their inability to pay high rents would be a recipe for an imbalanced city. Yet planners, politicians, and developers embrace or fail to question the assumption that the most appropriate use of a piece of real estate is whichever use will pay the most.

Such market deference was, of course, selectively applied in Battery Park City (as it is elsewhere). State officials used market ideology to justify why they were not willing to provide supports and subsidies for middle- or low-income housing on site, not to eliminate the considerable subsidies Battery Park City did enjoy. In fact, thanks to concerted state involvement in the design and financing of the neighborhood, tax subsidies, rent subsidies, and other benefits, Battery Park City reflected anything but free market economics.

The suburban strategy's (selective) deference to the market in determining land use endorses a related fashion in building public spaces. It has become stylish for governments to say that parks should "pay for themselves" with concession-stand fees, corporate sponsorship, and wealthy private benefactors.[2] Such a demand ensures abundant, luxurious park space in wealthy areas but offers no funds for parks in neighborhoods without the disposable income to support strings of cafés throughout a large park. In practice, this has often meant that parks are built as amenities for large-scale, high-end real estate development projects. Even in the case of upper-class neighborhoods, the greenspaces of Battery Park City "pay for themselves" only because residents' payments in lieu of taxes fund neighborhood parks rather than city services. Under such a regime parks cannot be planned in response to communities' needs.

The fact that Battery Park City is subsidized is not a criticism of it. To the contrary, it appears that subsidies are necessary, whether the intention is to build a high-rent neighborhood like Battery Park City or a low-income or working-class neighborhood. If building parks without public funding produces inequitably distributed parks, then subsidies are not the problem. They can be part of a just solution. Public sponsorship of parks frees the public to determine the best use for a space, or the best location for a park, without being hogtied by the requirement that the park generate its own revenue. The expectation that public goods can survive without public support is a libertarian fantasy: even in Texas, a state not known for its generous public financing, a report for the Department of Transportation could find no road in the state where gas taxes and tolls covered the cost of the road's maintenance.[3] Taxes are higher on gasoline than on most products, yet the report concluded that state gas taxes would have had to increase from 20 cents a gallon to as much as $5 *per gallon* for motorists to "pay their own way."[4] That is, just as parks, playgrounds, subways, buses, and trains are subsidized, so are highways and roads. The problem, therefore, comes not from the fact that Battery Park City was supported with tax money but that it represented a clear case of inequitable subsidy (in which subsidies flowed disproportionately to the wealthy). The suburban strategy quietly employs state subsidies for favored projects but argues against state support for disfavored neighborhoods, all the while inaccurately suggesting that such subsidies are mechanisms employed only on behalf of the "undeserving" poor.[5]

Battery Park City provides a vision of what can be achieved with subsidies for high-quality neighborhoods. State subsidies support neighborhoods and build usable, high-quality public spaces. These are things that everyone needs, and they should be provided equally to all.

There is one remaining puzzle to Battery Park City. How was such impressive state support marshaled during decades when state support for other more populist programs was being cut? Here Battery Park City is again instructive. The middle class and working class have been marginalized from the broader public political discourse just as they have been from Battery Park City itself. There was a time when an imagined "middle" held center stage in political discourse, from the local to the national level. Whether the middle was Nixon's imagined "silent majority" or the middle-class Queens residents whose anger undid John Lindsay's mayoralty, the middle was the rhetorical target of political promises. At the time, government funds in New York City built apartments for poor, working-class, and middle-class residents. (The Mitchell Lama program alone built nearly forty-five thousand middle-income units in New York City in the mid–twentieth century.)[6] Today, the legacy of the middle class's centrality lives on predominantly in empty signifiers, such as "middle-class tax cuts," which can be used to refer to policies benefiting a group nowhere near any middle.

The political disappearance of the "middle" (which in truth included both middle- and working-class people) is paralleled by the disappearance of the "middle" from the profile of who would live in Battery Park City. Such potential residents were pushed out, first by the financial executives like David Rockefeller who conceived of Battery Park City, then by progrowth politicians at the state and local level who endorsed the vision of Battery Park City as an executive enclave, and finally by the residents themselves.

The battle over West Street demonstrated the lengths to which residents went to ensure that the physical design of their neighborhood would continue to exclude outsiders. Through that battle, it was difficult to discern whom residents were seeking to keep out: the neighborhood was already relatively isolated and had little through traffic. While studies of residential exclusivity and residential segregation normally presume that residents are trying to exclude poor people or racial minorities, there were few poor, or African American, or Latino residents nearby. Nor did most residents justify the West Street barrier with racial code words such as "crime," "undesirables," or "bad neighbors."[7] Instead, they worried most about "tourists," people who had the means to visit Manhattan but lacked the cultural capital to fit in to upscale Battery Park City. Residents were likewise concerned about who would move in when rent subsidies were offered in the years immediately after September 11, even though rents would still have been out of reach of most New Yorkers. Battery Park City's social exclusivity focused not on the poor, who were socially and geographically distant from them, but on the middle class, who were the proximate group from which the upper class sought to distinguish itself.

In principle, it makes no difference who was to be kept out. But rhetorically, excluding the middle class is more contentious. In residential communities across the country, the middle class itself practices exclusivity: zoning against multifamily dwellings keeps renters out of the suburbs, gerrymandered school boundaries keep students of color out of predominantly white schools, discriminatory real estate practices help keep most U.S. neighborhoods highly segregated.[8] Were Battery Park City keeping out only the poor, the middle-class public would probably assent to such actions through "complicitous silence."[9] But as the sacred-if-eviscerated signifier at the center of political discourse, the middle class would likely take great offense to find a community designed to keep *them* out. In classic reciprocal fashion, the suburban strategy is both a product of the political marginalization of working-class and middle-class political constituencies that allowed projects like Battery Park City to be undertaken, and a means by which the middle class can then be spatially marginalized. The spatial marginalization of the middle class may be only an extension of other, no less noxious forms of segregation, but it has the political potential to turn a new constituency against such exclusivity.

At the same time, the suburban strategy has been employed to create public spaces in which there are precious few opportunities to object to suburban space's vibrant exclusivity. Use of most public spaces is highly local, so to the extent that the suburban strategy further spatially stratifies cities by race and class, people have less chance to observe uneven resource distribution and fewer shared public spaces in which planned or quotidian challenges to the inequities of the suburban strategy can be presented. In suburbanized public space, inequality between classes or races can become less evident.

Exclusion via the suburban strategy indicates the completion of large portions of the urban gentrification project. Jane Jacobs remains revered among urban planners, but the block-by-block diversity she celebrated is a relic that is being systematically dismantled. Such diversity is omitted from development plans by private developers or public agencies like the Battery Park City Authority. Such fine-grained diversity was necessary in the pedestrian city of more than a century ago. It was furthered by the integrationist demands of the civil rights movement and was given an unintentional boost during urban disinvestments in the postwar period, when people with less money could gain access to space in formerly more exclusive areas. But today there are no serious efforts to produce or preserve diversity in development efforts in New York, though there could be. As in plans for Battery Park City, diversity quickly falls victim to the suburban land use model that is hegemonic, as space is developed in large, economically homogenous zones.

In part because Battery Park City epitomizes such zones, the community appears to have recovered from the impact of September 11. Measured in the narrowest terms, Battery Park City is thriving. New apartment buildings have been constructed and occupied. In the ten years since the last census taken just before September 11, 2001, the community grew by 55 percent, from 7,951 to 12,390 people. In that decade, the median household income grew by 25 percent to $134,464. Today, over 35 percent of households bring in more than $200,000. The median home price—even *after* the 2008 crash of the nationwide housing bubble—grew by more than 120 percent from $316,000 to $703,000.[10]

But high salaries and home prices in Battery Park City are only one part of the story of the suburban strategy. Many observers of the city have noted that the suburban spaces of Manhattan are cleaner, prettier, safer, and more expensive than they have been in decades, and they take those signs as sufficient evidence of improvement. But a closer look at more important statistics reveals that application of the suburban strategy to New York has not strengthened the city. Instead, it has deteriorated the quality of life for most New Yorkers in measurable ways. The structural shifts in employment, away from unionized jobs and diverse manufacturing to jobs in the financial industry and in service industries, tourism, and other sectors that support primarily "casual" (part-time, low-wage, low-benefit) work, have impoverished the city. The emphasis on the financial sector and the implementation of filtered space and, later, suburban space, constituted the elite response to the 1975 fiscal crisis. While this strategy accounts for the rising fortunes of Battery Park City over the last ten years, it made life for most New Yorkers worse.

New York City's poverty rate was 14.5 percent in 1969, just 0.8 percent above the national average. In the 1970s, it rose rapidly to about 20 percent and has stayed there. Twenty-two percent of New Yorkers were below the poverty line in 1999, compared to 12.4 percent in the United States.[11] The city's middle class has shrunk, with increases in the small number of upper-income families and in the much larger number of low-income New Yorkers.[12] Researchers investigating possible causes of New York's persistently high poverty rate have concluded that the polarization of incomes brought about by the city's polarizing economy explains the increase.[13] The suburban strategy and the shift to the financial and service sectors of the economy have made more New Yorkers poor.

Likewise, unemployment has been a chronic problem in New York. The city's unemployment rate has been significantly higher than the national rate since the 1970s.[14] Unemployment is markedly higher among many groups of New Yorkers. Whether the economy is doing well or badly, for instance, African Americans typically experience unemployment rates double the national rate.[15]

In the recession following the housing market collapse of 2008, New York fared better than it could have. For two years its unemployment rate was within half a point of the U.S. figure.[16] But this concealed two more persistent problems. First, the rising numbers of the working poor meant that many New Yorkers could be working but still poor. Between 16 and 20 percent of New York families with children in which someone *was* employed were still below the poverty line.[17] Second, the city's long-term unemployment rate (those out of work for more than six months) has exceeded the national rate for more than twenty years.

Finally, since the 1970s, rising housing costs and falling wages have put New Yorkers in a desperate pinch. New York is a notoriously expensive place to live. But until the late 1970s it was not: housing costs relative to income in New York were comparable to other parts of the country. Only since then has housing become radically more expensive. Housing costs that are more than 30 percent of a household's income are deemed "unaffordable." In 2008, half of New Yorkers paid more than that. Nearly 30 percent were "severely burdened," paying more than half their income in rent.[18] The result of high rent burdens is not only less money for other needs and greater anxiety about paying the rent but higher risks of homelessness. Boosters of the suburban strategy have used the visual improvements to the center city and decreasing crime rates as evidence that the city has triumphed over the challenges that plagued it. But this appears true only from within the boundaries of suburban spaces like the Battery Park City citadel. Beyond those boundaries, expansive as they are, persistently high poverty, large ranks of the working poor, dogged unemployment, and the financial anxiety and instability of high housing costs pronounce the suburban strategy a failure.

That strategy remains contested, of course. Looking to Battery Park City, the reciprocal relation between space and social inequality ensures that public space will continue to be contested. While it may be true that Battery Park City, fully built out, is a historic artifact, opportunities remain. First, there will continue to be *re*development in Battery Park City and changes to the parks, streets, businesses, and buildings. Larger opportunities exist in the future as well: when the ninety-nine-year leases expire on the land on which the buildings are built, citizens will have another opportunity to demand more inclusive and affordable housing or more accessible spaces. The current glut of commercial real estate downtown begs for the development of a more diverse, sustainable economy that pays living wages. The most visible contested space is just across Battery Park City's boundary. The World Trade Center site is slowly being rebuilt. As of this writing, the steel frames of some of the buildings, rising from the foundation some sixty feet below, have just crested street level. But the ultimate use of those buildings and the public spaces around them remains a matter of intense debate. The varied potential uses of the memorial space likewise have yet to be considered.

Improving Public Space

Concerning the effects of space on community organizations, there was a fundamental dissonance between how close residents were to the Trade Center site physically and how marginal they felt they were to the planning process. This was particularly pronounced when Battery Park City residents felt other groups were being listened to more than they were. "I think when you're pushed to the edge of the process, even when you're an integral part of it—I think residents recognize that none of these people [families] will live there," said Manohar Kanuri. Clearly a site as significant as the World Trade Center needs to consider more than local needs alone. But residents remained angry that their needs were not being considered more seriously. Part of the problem stems from the World Trade Center site's difference, sociospatially, from most of the other sites residents and other people have compared it to, such as the Vietnam memorial in Washington, D.C., the *U.S.S. Arizona* in Pearl Harbor, and even the Oklahoma City bombing memorial, itself on the site of a former office building. The spaces these sites occupy are no longer part of people's everyday communities. Thus while those spaces could be used wholly for memorial purposes (or in fact for any number of purposes) without affecting nonusers, the World Trade Center's location in the densest part of the densest, most pedestrian city in the United States complicated the process immeasurably. Even if Maya Lin's design for the Vietnam memorial had been the flop critics first claimed it was, it wouldn't have made anyone's residential or work neighborhood dull, less livable, or less convenient, tucked away as it is in the memorial zone of the Washington Mall. But at the World Trade Center, it is possible that even a *well-designed* memorial could be poorly designed for neighborhood use. (Imagine if Lin's long wall, so effective in its setting, instead were inserted in Lower Manhattan, where it would block several intersections and interrupt residents from getting home and customers from getting to businesses.) The deadening effects of an inattentive or oversized urban redevelopment project as well as the incredible commercial value of the land are such that there are many more constituents than the memorial users themselves. In few other cases of memorial designs have there been arguments, as there were in Community Board meetings, between residents and families about how the memorial would destroy a neighborhood, or how droves of residents would "have to leave" if an element of the plan were built. (Some family members told residents to "just leave" their own neighborhood if they didn't like living next to the memorial.)

The most significant problems residents faced in the redevelopment process were produced neither by residents nor by victims' families. The redevelopment process is inevitably contested by a multitude of interest groups of widely varying influence. But building in this very public and publicly important site has not been helped by the faux-democratic approach of the Lower Manhattan Development Corporation and other governmental and quasi-governmental organizations. It would be challenging to work out people's differences under better circumstances, but the actual process made compromise far more difficult because residents, family members, and others had no assurance that successive compromises would not erode their claims entirely, or that they would be listened to at all if they were anything but loud and uncompromising.

Battery Park City provides indications of how people can make such projects more democratic and public. Maximizing public participation in decision making about public space is an excellent way to see that such spaces serve the broadest possible public need. In a very direct way, participation that is more public creates spaces that are more public.

My observations of a number of public space redevelopment proposals (for this project and for others) demonstrated again and again the positive role of public involvement in planning. Once residents moved into Battery Park City, each subsequent building project became a contest between the Authority's vision (based on the master plan) and divergent needs of the residents. Over the decades, residents demanded changes to the plans —more active recreation space, more playgrounds, dog runs, baseball diamonds, community gardens. The plans were clearly improved as a result.

The mechanism by which public participation was incorporated has varied. But what public space planning processes consistently show is that a small number of decision makers, designers, or developers can design a space only for a small number of uses. In contrast, planning processes in which the needs of a broader segment of the public are legitimized, solicited, and addressed produce parks and public spaces that are more widely used, by a larger number of people engaged in a broader array of activities, and are therefore more public.

The value of incorporating public input in public space planning is demonstrated in a comparison between responses in privately controlled spaces and publicly controlled spaces. In privately owned public spaces, the problem of dogs is dealt with, in top-down fashion, by banning dogs. In public parks (as in Battery Park City), dog owners' needs are legitimate, so the problem of dogs running off leash is addressed by creating "dog runs"— designated, fenced-in areas just for dogs, with amenities like canine water fountains, easily cleanable and more sanitary surfaces, even separate areas for big and small dogs. The result is a park that is more suitable both for dog owners and non–dog owners alike. Examples of legitimating uses that at first seem nonconforming go farther. In one extreme case in New York, Steven Brill was repeatedly ticketed and arrested for destroying plants in

Central Park. In fact, he was leading tours of the park, sampling and showing off the edible plants that grew wild there. After he convinced the Parks Department that his tours were not damaging the flora (which, like a tenacious blackberry bush, quickly grew back), his interests were legitimized. Since then, he has led permitted tours of several city parks on a regular basis, including tours to hundreds of public schoolchildren every year.[19] As I found, the plants' surprising flavors cannot help but change the way participants see what is growing around them. Because the Parks Department has a structural and cultural commitment to recognizing and validating a wide range of public uses (criticisms of the Parks Department notwithstanding), it is a fair representation of a relatively democratic process of decision making regarding the permissible uses of public space.

Public input contributes directly to a lively public space because political processes communicate the otherwise unknowable needs of diverse groups of interest holders to those who will make decisions about the dispensation of resources, space, or subsidies.[20] Parents request a playground, vendors lobby for permission to sell food, basketball players ask for the courts to be refurbished. Compared to the top-down style of a planner-demigod, democratic processes do a better job of filtering up people's needs, and to the extent that a society is democratic those needs are recognized as legitimate and are incorporated.

Such public participation is sometimes the source of frustration among observers and participants alike. In the ongoing contest between Jane Jacobs's and Robert Moses' visions of urban planning, there was a brief resuscitation of Robert Moses in the opening years of the twenty-first century. At the time, commentators admired the grand public works constructed under Moses and asked whether it would not be preferable to have one powerful actor such as Moses, able to move mountains and redirect highways through residential neighborhoods.[21] The ritualized public input processes that characterizes contemporary land use planning overwhelmed Lower Manhattan during the endless rounds of hearings about West Street, bus parking, and memorial plans. In criticizing those protracted public input processes, commentators pined for the days when someone like Moses could steamroll neighborhood objections and get things built.

Such misplaced nostalgia for Moses misunderstands the complexity of cities and the purpose of democracy. Cities' complexity was well illustrated at a meeting regarding West Street. For some time, plans had indicated that an underground pedestrian walkway would connect the subterranean shopping mall on the rebuilt World Trade Center site to the World

Financial Center in Battery Park City. When state officials unveiled large side-view drawings of the walkway, one resident asked the most obvious question: Why had the walkway been designed so that pedestrians had to walk down three steps at the beginning of the corridor, across the passageway, and then up three steps when they reached the other side? He understood why old buildings might have stairs that obstructed wheelchairs and strollers. But shouldn't a new tunnel, he asked, be level all the way across? The state representative nodded sympathetically. The steps were there, he explained, because beneath West Street the ground was so clogged with electrical wires, gas mains, hundred-year-old train tunnels, river water intake pipes, and other utilities that designers had only about six inches of leeway regarding where to place the tunnel. That tight a constraint meant that the only way to fit the walkway under the street was to design it so that it descended three steps, barely squeezed between some of those pipes, and rose back up again at the other end. Planners had no room to raise the passageway three steps. In a city as dense as New York, even underground one can hardly move an inch without bumping into someone else's uses of space.

While the pipes and conduits provided hard, visible evidence of the density of demands placed on a single space, individuals' uses of space are just as tightly packed. Thus any major redevelopment project digs through ground that is thick with the conduits of people's lives in public space: vendors' livelihoods, children's play areas, commuters' routes, parents' safety concerns. In such situations, one important function of democratic processes is to identify, legitimate, and reconcile such competing interests. As the political theorist Benjamin Barber defines it, "strong" democracy "resolves conflicts . . . through a participatory process of ongoing, proximate self-legislation and the creation of a political community capable of transforming . . . private interests into public goods."[22] Democratic processes like those used in development projects (though those processes are typically too weak for the needs of the public to challenge developers' priorities) are necessary to identify what a particular project must accomplish, what the needs of a community are, and what form a new public space should take. Such solicitation of input is more time consuming than the Moses model, but it is irreplaceable for creating complex spaces that can serve the varied needs of a large population.[23]

Processes that force designers to accommodate as many of those concerns as possible produce better, more functional, richer and more meaningful spaces. Participatory processes have been no small part of the secret to Battery Park City's evident success. The original master plan called for undifferentiated, passive outdoor space. For decades, residents have doggedly demanded, at meeting after meeting, the development's inclusion of a wider range of needs, from volleyball courts to marinas to safety fences, providing Battery Park City a more usable landscape and more significant collective memory than ever envisioned in the 1979 master plan.

In this way, community participation has contributed some of Jane Jacobs's finely woven urban fabric to what would otherwise have been a simplistic, brand-new neighborhood plan. The Community Board has played a valuable role in this process, both as a forum for residents and as an advocate in its own right. But shifting from the exclusivity of suburban space to more public space requires processes more inclusive than community boards and larger than Jane Jacobs's block. Community boards are the wrong scale at which to judge development projects as significant as the World Trade Center or Battery Park City, because the members represent only the interests of the local neighborhood (and the typically prodevelopment politicians who appointed them). What would a "community board" consisting of members from Brooklyn, Queens, Staten Island, and the Bronx as well as Lower Manhattan have said about the massive state subsidies when they were first planned for Battery Park City? It is not hard to imagine members from other neighborhoods objecting to such largesse being handed out to an already privileged neighborhood before other pressing needs were addressed. The Battery Park City proposal would have had to compete with proposals for other uses of the money and the land. Thus, while community-level debate of truly community-level decisions —like applications for liquor licenses, small building variances, and individual buildings—may be appropriate, large projects that will use millions of state dollars over decades need to be debated by a broader distribution of the public.[24]

A more inclusive system of decision making about place building would counteract some of the inequity and exclusivity of future plans for projects like Battery Park City. A broader debate would probably result in demands that the resources and the space provide more public benefit. Discussions regarding World Trade Center redevelopment have been extensive over the past ten years, but the process has not been structured in a way that forces the city to seriously consider such input. The necessary diversity of spaces—from intimate community space to large, inclusive public space —requires broad public input.

Jacobs's complexity is needed in Moses-style projects. But Jacobs's vision for the city is not sufficient either. First, Jacobs underplayed the influence of powerful economic actors in shaping the city in ways that were beneficial for them but deleterious for the rest of us. Second, by all available evidence, Jacobs's recommendations for urban planners celebrated integrated diversity but did not provide adequate means of maintaining a diverse neighborhood and staving off the kind of homogenizing gentrification and racial resegregation that destroyed diversity in almost every neighborhood she celebrated. Rectifying the effects of the suburban strategy requires counteracting the influence of powerful economic actors (such as structurally speculating real estate developers) and making a firm commitment to racial and economic integration in cities. While greater participation stakes out a role for the public in improving suburbanized spaces, achieving the dual goals of contesting capitalism's power over the landscape and integrating all participants in the city requires a role not only for the public but the state as well.

Under the presumption of the suburban strategy, government not only abdicates commitments to intervene directly to improve the lives of low-income people but now denies the singular role it can play to overcome chronically high levels of economic and racial segregation. State commitment to integration is necessary. Particularly given the scale of segregation that we confront on the American landscape today, no development project is justifiable if it does not take considerable strides to counter segregation. It is not legitimate for the state to endorse plans, like those of Battery Park City, that are not designed to be of use to the whole city. Public input may be enough to create designs for diverse activities in public space, but state protection of the principles of equal protection, equal access, nondiscrimination, and integration is required to create spaces that are used by residents of every race and class.

In these ways, the conflicting visions of Moses and Jacobs could be resolved. Battery Park City, as it was built, was a partial realization of Jacobs and Moses but left out Moses's authoritarian populism and Jacobs's grassroots populism. What would a project look like that did include those two aspects? Such a plan would make the public spaces true destination spaces that attracted New Yorkers from across the city, while creating residential areas that were diverse, integrated places rendered stable and secure through eyes on the street and strong social networks. Greater participation by people who would use a space would provide the necessary intricacies to counter the broad brushstrokes of Moses-style megaprojects and make communities usable. Meanwhile, decisive and powerful state commitments on behalf of economic and racial integration would counter the exclusivity that can often stem from decisions made by those smaller groups. More democratic input ensures that residents' needs will be met, while state requirements that all people benefit and participate in such projects counters the exclusivity that can infect the reciprocal relationship that shapes cities.

Battery Park City offers much to the city, both faults and assets. It is a model of large-scale planning that proves such projects can create livable, vital communities. It is a reminder that designs intended to narrowly benefit an elite sector have damaging ripple effects, as those designs reciprocally reinforce exclusivity among subsequent generations of residents. But Battery Park City has always been engaged in the kinds of committed, messy, public debates that can make such projects more public, and more just.

Appendix A

"*September 11, 2001*"
Reprinted from the *Battery Park City Broadsheet*

(The following account of September 11, 2001, was published in the *Battery Park City Broadsheet*, the community newspaper published by Alison and Robert Simko out of Gateway Plaza. This excerpt provides a peerless account of the collapse of the World Trade Center Towers on September 11, 2001, from the perspective of the people who lived next door in Battery Park City. It was published September 22, 2001, in the first edition of the Broadsheet after the destruction of the Trade Center.)

Robert Simko/The Battery Park City Broadsheet © 2010

Shaken Lower Manhattan Neighbors Find Each Other after September 11 Disaster; Vow to Stay Together and Rebuild the Community

Battery Park City Authority, Working with Mayor's Office of Emergency Management, Pulls Together a Huge Effort to Restore Evacuated Neighborhood; First Buildings Reopen September 20

BPCA Looks to Quickly Develop 2 Million Square Feet of Site 26 to Help Replace Lost Office Space

8 a.m., Tuesday morning, Primary Day September 11. It's the familiar rushed jumble of breakfast, dressing, plans for the afternoon, making two lunches, tucking them into backpacks. Waiting for the elevator, Lucy describes a bad dream involving explosions and fire. Fifteen minutes later, Theo and Mom head off for school by bicycle, dodging last-minute campaigners along the way. It's a beautiful, sunny morning, and the world is full of friendly hellos and smiles. And suddenly we are fleeing smoke and destruction, shaking, crying. And where is the children's father?!

September 11, 9:03 a.m. The second plane pierces the South Tower, raining debris below. The people scream, and suddenly and brutally recognize the shift from tragic accident to state of siege.

After facing unspeakable terror on September 11 when two planes hit the World Trade Center and caused their fiery collapse, the residents of lower Manhattan are coming together, determined to rebuild their community. Robert Simko, Broadsheet publisher, eventually checked in unscathed, at least physically, but some neighbors are not accounted for. What follows is a range of voices recounting experiences during the week of September 11 through 18.

Tuesday, September 11

8:48 a.m.—Flying south over lower Manhattan with a terrible roar, a Boeing 767 slams into the North Tower of the World Trade Center. "I was standing in front of P.S. 234, greeting all the voters and parents and just feeling the energy of the first day of kindergarten—and I heard this groaning whine of a jet and looked up to see it overhead so close as though you could almost touch it," says Cass Collins. "It sailed dead-on into the tower, and it was clear to me in that instant that it was deliberate. And then I saw this inverted V-shaped scar on the tower begin to glow inside."

8:55 a.m.—BPC attorney Les Jacobowitz sees people engulfed in flames on the sidewalk at the base of the World Trade Center. As construction workers help them, Mr. Jacobowitz runs to the intersection of West and Vesey Streets to direct traffic away from the disaster site.

9 a.m.—Federal agents commandeer telephones at P.S./I.S. 89. Principals at downtown schools try to calm panicked parents and children.

9:03 a.m.—Flying over New York Harbor and southern Battery Park City, a second Boeing 767 hits the south tower. Eric Nevin is watching from his apartment at 71 Broadway. "The south tower is framed in my window

September 11, 9:30 a.m. Part of the plane's landing gear is roped off at the southeast corner of Albany and West Streets.

... and as I stared at it, the second plane appeared and impossibly smashed into it, creating a horrible fireball explosion."

Sterling Rome is watching from his window at 166 Broadway, directly across from the World Trade Center. "I was thrown backwards to the floor and could actually feel the heat of the explosion through the glass. A fireball rolled down the tower and then spewed debris everywhere," he wrote for CNSNews.com. "Parts of the building and the plane fell down like rain. I could hear the wailing of the people in the street below as they fled."

Throughout lower Manhattan, residents, workers, and candidates freeze in horror. People react in different ways: some weep, some tremble, some excitedly take photographs, some slump and stare at the sidewalk, some vomit, some make nervous jokes.

9:20 a.m.—Phil the windowwasher sees eight people jump in a group from an upper floor of the World Trade Center. Helicopters land on the BPC ball fields. Fireboats pull up to the BPC esplanade and begin pumping seawater through hoses attached to fire trucks parked near West Street.

9:59 a.m.—The first tower collapses. "I still have trouble believing, understanding," recalls Mr. Nevin. "The south tower issued a sickening rumble like that of an avalanche, and my eyes took in the beginning of the collapse, the impossible accordion-like surrender of this tower-mountain that had always seemed among the most permanent of human-made structures."

Mr. Rome writes, "I took a step towards the windows and saw the south tower start to fold. It was like a horrible, terrible dream.

Everything seemed to move in slow motion. The tower gave way a little at first, then all at once. People were still leaping from the building as it fell. My wife was next to me now, gaping. We watched as the south tower came down right in front of our eyes. Tower fragments the size of a football field fell in every direction. There was a billowing black cloud of ash, smoke, and debris forty stories high that rushed through just before impact. I embraced my wife and we watched and waited to die."

All over lower Manhattan people flee, abandoning baby strollers, dropping briefcases, purses, bags, running right out of their shoes.

Wives are separated from husbands, children from parents. But the Galloway family is together in the Gateway Plaza ring road. "We ran to the river through the back of our complex," says Jeff Galloway. "I looked back and saw this black, fiery avalanche heading straight at us. It was like an Indiana Jones movie, except I knew we couldn't outrun this thing. I thought, 'this is it—the fire will be here in a second, there won't be any oxygen to breathe,'

September 11, 10:00 a.m. Ash, smoke, and debris from the tower's collapse curl around the World Financial Center and Gateway Plaza and advance down South End Avenue. (Photo by Bill Hartford)

when it hit. Total blackness. We couldn't see anything, our hands, our kids, the dog, each other—nothing. I held tight to Liam . . . and reached out my hand and found Paula and Kiera and the dog, and we all just held tight there."

"Running uptown maybe a half block from our building, I 'knew' a third plane was about to crash into the towers. Wrong," remembers Kathleen Bachand. "The noise was the roar from the first tower crumbling and the ground going out from under us. Everyone was screaming to get down. We were covered by debris and choking. I was holding onto Joe when the billow of gray smoke came over us and went to pitch black. I couldn't see him and I was holding him. When it started lifting just enough to see through the gray we started running again."

"I was in front of St. Paul's Chapel," Kathryn Soman tells the *Broadsheet.* "I felt confident that people were getting out. The towers were burning above me and it never occurred to me they would come down. But then there was this terrible humming noise, and one corner of the building started to bubble, and then it came down and this cloud of dust and debris engulfed us. You could identify glass, metal, electrical wire—and I was completely aware that mixed in with all this stuff were people; from the ashiness, I knew that I was breathing in the atoms of people."

10:28 a.m. — The second tower collapses. From his apartment window in the Gateway Plaza 600 building, Udayan Gupta, a financial writer, sees the tower fall and the dark cloud roil toward him. Suddenly, his windows burst in and shatter. He runs to the basement.

Journalist Stacey Cahn is trying to file reports from her 30th floor apartment facing the World Trade Center: "We didn't know what would come along in those plumes of smoke headed our way. We ducked. We froze. We prayed. After the broadcast antenna came down (and it seemed to fall in slow motion, as if to bid us adieu), we lost all power. We couldn't believe what was happening to our neighborhood, considered so safe after the 1993 bombing — and what this meant for all of us."

10:45 a.m. — Andie Chester runs out of the D.J. Knight office on West Thames Street where she has been under a desk with her son Dave and his friend Pierce. A police officer herds her back into the office, saying a suspicious package is lying near the Museum of Jewish Heritage.

11 a.m. — "I'm running with my baby near the Museum of Jewish Heritage and a man opens his car door and calls to me," says Dean Lubnick. "I say, 'take my baby. I have to find my wife.' He says, 'No, your baby needs you. I'll find your wife.'" The hero, David Chase, catering manager of the Ritz-Carlton, finds Mr. Lubnick's wife, Lee, and leads her back to her husband and baby Rebekah.

11:30 a.m. — Tugboats and other vessels push up to the esplanade and evacuate hundreds of shaken people, some injured. Denise Cordivano, co-director of the Battery Park City Day Nursery, escorts children to the boats whose parents did not pick them up.

At noon, Glenn Fennelly, a pediatrician who lives at Parc Place, races downtown by ambulance from Jacoby Hospital in the Bronx. He knows his wife Elisa and daughter Yuri have been evacuated. At South Ferry, he treats people with breathing problems and dislocated shoulders, but soon, there's nothing for him to do. A group of two dozen doctors from NYU Downtown Hospital are turned away. Mr. Fennelly walks to Battery Park City, but a policeman bars him from entering his building, saying looters have been arrested.

1 p.m. — Mary White is in the lobby of the Gateway Plaza 200 building.

"We thought there was a gas leak. We thought the building was going to blow. It was mass panic. Some people were screaming to get out, some wanted to go to the basement."

3:30 p.m.—Some of the last evacuees by boat cross the Hudson to welcoming arms in Jersey City.

7:30 p.m.—"As night fell, we lit candles in the windows and noticed a handful of other neighbors following suit," says Ms. Cahn. "We wanted someone to know we were still there, and we wanted others to know they weren't alone."

Wednesday, September 12

7 a.m.—Eerily quiet except for a wail of sirens. The air downtown is smoky. The barricades have been moved from Franklin Street to Chambers Street. Sixty men stride down Hudson Street wearing hardhats.

Journalists from all over the world have arrived and set up camp. At her apartment in Gateway Plaza, Georgie Meinhofer, an elderly woman who lives alone, begins to make breakfast for a few neighbors who have stayed. When the knock comes on her door to evacuate, she invites the officers in for some pierogis.

The Battery Park City Authority puts up a message board at its web site *batteryparkcity.org*. James Gill, BPCA chairman, makes a statement to the world: "As we slowly recover from the horror of the World Trade Center attack, the Battery Park City Authority is profoundly grateful to report that not a single person was killed or injured on Battery Park City property . . . We are digging ourselves out. We are engaged in comprehensive cleanup operations and will repair and rebuild wherever necessary. In this way we will demonstrate to those who perpetrated these vicious, dastardly, and cowardly attacks that we have prevailed and they have lost."

Armed National Guardsmen park their humvee at the intersection of Broadway and Walker Street. It's a show of force, a guard explains, to scare away would-be looters.

Having talked their way past police officers, sandal-clad residents walk through dust and rubble to their dark apartment buildings, passing groups of workers in hazard gear, head to toe.

Thursday, September 13

Restrictions and security tighten around Battery Park City as the FBI begins to comb the area for evidence. The Federal Emergency Management Agency declares lower Manhattan a disaster area and Gateway Plaza is designated a crime scene by the FBI, whose agents have broken down doors in the 600 building in search of evidence. Residents are escorted to their apartments only to retrieve pets and medication.

September 15, 5:30 p.m. Men in uniform at ease near Hanover Square outside Delmonico's.

Friday, September 14

People walking on the mostly deserted streets of the financial district pause at a window display at 110 Maiden Lane. It's the home of the Skyscraper Museum, and one of the main subjects of the display no longer exists.

Saturday, September 15

Residents of lower Manhattan stream into a basketball court at the corner of Canal and Thompson Streets, drawn to a meeting called by downtown leaders and publicized by word of mouth and the Internet.

Approximately 500 people gather, most seeing each other for the first time since Tuesday. Tears and hugs abound.

10:15 a.m.—Bob Townley, executive director of Manhattan Youth, calls the meeting to order: "The lower Manhattan community stands ready to assist in the rescue effort of our neighbors, co-workers, fathers, mothers, brothers, and sisters. No matter what it takes. We stand ready to assist those who have lost their family members. As a community, we stand ready to rebuild commerce, businesses, schools, parks, and homes. We stand ready to help our new homeless and jobless. We stand ready to return to normal. Gain access to our homes. Open our schools. Socialize our youth. Open our Little Leagues, recreation programs, and care for our seniors. Our children are scared. We understand. We tell them to be brave. We will help them thrive. We will not leave. We will rebuild our community."

At Pier 40 and at the Bowling Green subway station entrance at Battery Park, people press against the barricades and beg the National Guard to let them get to their apartments. Sergeant Ted Andrysiak wins the hearts of the crowd even as he continues to deny most of them access to Battery Park City. He is unflappable and understanding, and he represents the quiet heroism of the thousands of emergency workers in lower Manhattan.

"A residential neighborhood, especially one that was so vibrant like ours, will cease to exist if you don't let us back in," warns Jeff Galloway.

September 11, 3:30 p.m. Charles and Uriah Frederick walk out of Gateway Plaza, past the school bus stop and taxi stand in front of the 600 building, which sustained the most damage from the World Trade Center collapse.

Appendix B

Methods

I had conducted observations of the public spaces of Battery Park City and the World Trade Center in the spring of 1999 and in August 2001 and had presented that research the month before the World Trade Center was destroyed. I returned to Battery Park City in January 2002 and conducted the bulk of my ethnographic research between then and April 2005, focusing on the area's public spaces, meetings of community organizations, and neighborhood events. I used three primary forms of field observation: attendance at formal meetings, attendance at community events, and visits in which I was a participant observer in the unplanned life of the public spaces. To obtain a sense of the neighborhood independent of residents' representations of it, I also conducted less socially engaged observations using repeated observations and quantitative measurements of selected spaces to establish, for instance, who used the public spaces, when, and how.

As part of my observations, I attended two years of Community Board 1 meetings, going to the meetings of several of the subcommittees whenever they might be addressing topics relevant to West Street and community redevelopment projects generally. I also attended the public hearings of the Lower Manhattan Development Corporation (LMDC) and the New York State Department of Transportation (DOT) that were part of environmental impact statements for the West Street project, as well as other LMDC meetings, where Battery Park City residents voiced their opposition to projects like the tunnel plan discussed in chapter 7. I attended the occasional meetings and events of community groups, but community groups often had few physical meetings and did much of their work online, which I followed through e-mail announcements, bulletin board archives, and e-mail lists. In some cases members of the organizations gave me the group's earlier correspondence or some of their own research, and the manager of an online bulletin board provided me with archives of posts immediately after September 11.

Interviews, particularly of residents but also of other key players in Battery Park City, provided rich articulation of residents' views of their community, neighbors, and values. I interviewed thirty residents, often more than once, as well as others, including Battery Park City Authority employees, private and city-employed urban planners who worked on Battery Park City, and activists from outside the neighborhood who had worked on Battery Park City issues. Respondents were evenly split between men and women and ranged in age from thirty to seventy. Proportional to the two largest racial groups in Battery Park City, more than 15 percent were Asian and the rest were white. Most residents I interviewed had lived in Battery Park City during September 11 and moved back afterwards, though I interviewed some who had left because of September 11 and some who had arrived later, to see if there were differences of opinions between those groups and long-term residents. The choice of interview subjects was biased toward those who were active in Battery Park City's public, political, and community life, particularly those active in the debates over redevelopment. This group was heavily weighted toward those who had been in Battery Park City for five years or more. About 20 percent of my respondents were less active members of the community whom I sought out using networks unconnected to contacts I had developed in Battery Park City, to compare their views to those of people who were more publicly visible.

Finally, I benefited immensely from several other sources: back issues of the local paper, the *Battery Park City Broadsheet*, offered not just an overview of community activity but a means to compare activity before and after September 11. Several opinion polls conducted during my research offered quantitative points of comparison to my qualitative findings regarding residents' attitudes.[1]

Clearly, residents' extended debates over the redevelopment of Battery Park City and the World Trade Center represent an exceptional period in the community's history. But previous research I had done had shown me that understanding the effects of public space and design required examining a community at exceptional moments: most of the time space is a silent social structure, shaping the social landscape without being noticed by those who inhabit it. Only at moments such as this one, when city officials proposed reshaping the space, are residents forced to identify their preferences for the space, the value a design holds for them, and the connection between the spatial design of the neighborhood and their definition of the community.

While this period is in some ways exceptional because redevelopment proposals fostered resident mobilization, activism is always periodic.[2] Battery Park City residents' mobilization in this period is comparable to previous activism there to preserve ball fields, mitigate large construction projects, influence park design, and relocate the ferry landing. Further, much of the activism I studied was based in groups formed before September 11. Thus resident activism in response to these redevelopment proposals was similar to that addressing earlier issues and provided an accurate indication of Battery Park City's community activism at the high point of a cycle.

Using Real Names

A shift away from anonymity has been under way in sociological ethnography in recent years. Whereas all ethnographies used to change the names of respondents and even alter (albeit unconvincingly) the names of the cities in which the studies were set, many studies today are published using the real names of participants. Researchers are showing the Institutional Review Boards that oversee their work that changing people's names is neither necessary nor sufficient for conducting ethical research. There are of course situations in which names must be disguised to protect anonymity, for instance when respondents discuss illicit activities such as drug dealing or civil disobedience plans, or when their participation could threaten their job or other relationships. But in many cases the need is less evident. Mitchell Duneier has made the most compelling case for using real names in *Sidewalk:* "To disclose the place and names of the people I have written about holds me up to a higher standard of evidence. Scholars and journalists may speak with these people, visit the site I have studied, or replicate aspects of my study. So my professional reputation depends on competent description."[3]

I adopted Duneier's convention of using real names. Though I felt strongly that real names should be used, in actuality I had no choice with many of my respondents. There was no disguising the terrorist attack across the street, or disguising the only state-sponsored luxury project nearby. From that point, it would not be difficult to identify the public figures I was studying: the head of the local Community Board, the editor of the local paper, the founder of a prominent opposition group. Doing so, more importantly, kept me honest, and I noted through the process of writing up my research that at several junctures writing with the knowledge that people's real names would be attached to their statements forced me to give those statements in a fuller context and to address their views more seriously and more respectfully than I could have had I simply sought anonymous quotes to support the prefabricated expectations provided me by the existent literature on citadels. In that respect I am confident that the use of real names made my research better.

Both Duneier and William F. Whyte (in a later edition of Whyte's *Street Corner Society* that identified the neighborhood and the real names of the primary participants) noted that their informants generally didn't care much about anonymity. People from the research sites rarely actually read the book, so respondents' identities were kept unknown whether they wanted this or not.[4]

Rather than protecting informants, as Duneier writes, "when I have asked myself whom I am protecting by refusing to disclose the names, the answer has always been me."[5] As someone who did earlier research with great anxiety about showing people what I had written about them, this observation of Duneier's has always resonated with me. Hiding the author from respondents has indeed seemed a tempting feature of using pseudonyms.

That said, the tendency to use real names in ethnographic research has become established enough that I feel comfortable also offering a caution. In looking at my own work and the work of others, I suspect there is a degree of analytical rigor more easily achieved with pseudonyms than with real names. Analyzing a situation, in the way that is one of the fundamental distinctions between ethnography and journalism, happens more readily when respondents have been partially abstracted. Part of what changing a name from Philadelphia to "Eastern City" signifies is that the conclusions about race relations from that study are not simply the idiosyncratic findings of a particular place but more abstract truths generalizable to the race relations of a range of eastern cities, or even the country as a whole. As I noted in the introduction, writing critically about people one becomes close to is difficult, and a critical stance is easier to maintain against more distant, abstracted actors than against closer acquaintances. In the end, I believe I gained more by using real names than I lost, but there is a cost to either choice, between the verisimilitude and honest representation of real names and the trenchant criticism and rhetorical power that can be unleashed against people hidden behind pseudonyms.

There are of course cases when anonymity is preferable or necessary. I certainly hope I have adequately protected my informants, and I have not used real names whenever a respondent expressed that preference (I gave them an explicit choice) or when I judged it more judicious not to use a name. But by appearing in these pages as who they are, the residents of Battery Park City present themselves more honestly and provide, I hope, a more unvarnished truth.

Notes

Notes to the Introduction

1. Authors have regularly cited Battery Park City as an example of a global city citadel. M. Christine Boyer, "Cities for Sale: Merchandising History at South Street Seaport," in *Variations on a Theme Park: The New American City and the End of Public Space*, ed. Michael Sorkin (New York: Hill and Wang, 1992); M. Christine Boyer, *The City of Collective Memory: The Historical Imagery and Architectural Entertainments* (Cambridge, MA: MIT Press, 1994); Richard Sennett, *The Conscience of the Eye: The Design and Social Life of Cities* (New York: W. W. Norton, 1990); Francis Russell, "Battery Park City: An American Dream of Urbanism," in *Design Review: Challenging Urban Aesthetic Control*, ed. Brenda Case Scheer and Wolfgang F. E. Preiser (New York: Chapman and Hall, 1994), 197–209; Sigurd Grava, "Battery Park City in the Last Decade of the Twentieth Century," in *Between Edge and Fabric: Battery Park City*, exh. cat., ed. R. Strickland (New York: Columbia University, Graduate School of Architecture, Planning and Preservation, 1991), accompanying the exhibit *Battery Park City Imagined*, World Financial Center, July–September 1991; Luc Nadal, "Discourse of Urban Public Space, USA, 1960–1995: A Historical Critique" (PhD diss., Columbia University, 2000); Phillip Lopate, "Planners' Dilemma," *7 Days*, February 15, 1989, quoted in Susan J. Fainstein, *The City Builders: Property, Politics, and Planning in London and New York* (Cambridge, MA: Blackwell, 1994), 183.

On citadels in global cities, see John Friedmann and Goetz Wolff, "World City Formation: An Agenda for Research and Action," *International Journal of Urban and Regional Research* 6, no. 3 (1982): 309–44; Janet L. Abu-Lughod, *New York, Chicago, Los Angeles: America's Global Cities* (Minneapolis: University of Minnesota Press, 1999); Peter Marcuse, "The Enclave, the Citadel, and the Ghetto: What Has Changed in the Post-Fordist U.S. City," *Urban Affairs Review* 33, no. 2 (1997): 228–24.

2. Raymond W. Gastil, *Beyond the Edge: New York's New Waterfront* (New York: Princeton Architectural Press, 2002), 46.

3. *Suburban strategy* is not meant to imply the cultural slight sometimes implied by *suburban*; rather, it refers to the postwar model in which segregation is maintained by using physical distances (rather than gates or covenants) to separate different groups, and large residential neighborhoods are created that are relatively

homogenous by class and race. In other ways, as this chapter makes clear, Battery Park City and New York remain distinctly urban.

4. For some of the earliest academic assessments of September 11, see Michael Sorkin and Sharon Zukin, eds., *After the World Trade Center: Rethinking New York City* (New York: Routledge, 2002).

5. The solar panels of the Solaire high-rise not withstanding, Battery Park City's greatest environmental contribution was that only 5 percent of residents commuted alone by car, while their counterparts who lived outside the city often logged many miles in the heaviest of sport utility vehicles, in a region with some of the longest commute times in the nation. An unparalleled 37.8 percent walked to work.

6. Data come from the 2000 U.S. Census and from census reports generated on socialexplorer.com.

7. On graduate degrees, see Philip Kasinitz, Gregory Smithsimon, and Binh Pok, "Disaster at the Doorstep: Battery Park City and Tribeca Respond to the Events of 9/11," in *Wounded City: The Social Impact of 9/11*, ed. Nancy Foner (New York: Russell Sage, 2003), 79–105. Income statistics are from the 2000 U.S. Census.

8. Eduardo Bonilla-Silva, *Racism without Racists: Color-Blind Racism and the Persistence of Racial Inequality in the United States*, 2nd ed. (New York: Rowman and Littlefield, 2006).

9. 2000 U.S. Census. My thanks to William Kornblum for census data on race, gender, and many other factors. Conveniently, Battery Park City is contiguous with Census Tract 317.01.

10. The 2 percent figure includes those who identified themselves as African American "alone" and "respondents who checked multiple boxes." The proportion of the U.S. population at the time that was African American was 12.9 percent. "QT-P5. Race Alone or in Combination: 2000," U.S. Census, www.census.gov, accessed July 2, 2004.

11. Measuring exactly how many people moved out after September 11 has been difficult, and obtaining a precise number will never be possible. The *New York Times* reported in 2003 that "the Census Bureau says that it is too soon to count how many people moved away" (Eric Lipton and Mike McIntire, "Two Years Later; Jobs, Tourists and Nail-Biters: Taking the City's Pulse after 9/11," *New York Times*, September 11, 2003). However, a range of sources corroborate residents' estimates that about half moved away. The Battery Park Synagogue lost at least half its members, according to the cantor and spiritual leader. A survey by the Alliance for Downtown New York found apartment vacancies in Lower Manhattan of 50 to 70 percent in February 2002 (Dennis Hevesi, "Residential Real Estate; Downtown Rentals Up; Survey Cites Subsidy," *New York Times*, June 14, 2002). Vacancies in Battery Park City were at least 25 percent for rentals and condominiums that February, according to building owners and realtors. If 25 percent were vacant at that point, half of residents could have moved out over the course of two years (Alan S.

Oser, "Deciding Whether to Make a Home Near Ground Zero," *New York Times*, February 24, 2002). One year after September 11, 2001, a more comprehensive survey of twenty-seven residential buildings in Battery Park City and other blocks around the World Trade Center found vacancy rates of up to 75 percent. Battery Park City buildings presented in an accompanying visual graphic all appeared to be at least 25 percent vacant, with most around 50 percent (Greg Winter and Kevin Flynn, "Living Space: About the Survey," *New York Times*, September 11, 2002). Roughly 700 of Gateway's 1,712 units were vacant after the complex reopened, and a local school expected a 28 percent drop in enrollment as the first day of school approached a year later (Greg Winter, "Reverse Exodus in Apartments Near Tower Site," *New York Times*, August 20, 2002).

12. In June 2007, two-bedroom apartments were being rented for an average of almost $6,000 a month, far above the total monthly *income* of most New York households. Real Estate Group, "The Manhattan Rental Market Report," January 2008, www.tregny.com.

13. On New York's twentieth-century working-class history, see Joshua Freeman, *Working-Class New York: Life and Labor since World War II* (New York: Free Press, 2000). On how the financial, insurance, and real estate industries came to dominate New York, see Robert Fitch, *The Assassination of New York* (New York: Verso, 1993).

14. Friedmann and Wolff, "World City Formation," 325.

15. Saskia Sassen, *The Global City: New York, London, Tokyo* (Princeton: Princeton University Press, 1991), 4, 251; Manuel Castells, "European Cities, the Informational Society, and the Global Economy," *New Left Review* 204 (March/April 1994): 26. Although there are conceptual differences between world cities and global cities, the focus here on citadels renders those differences tangential to the present discussion. Mike Davis, *City of Quartz: Excavating the Future in Los Angeles* (New York: Verso, 1990).

16. Sharon Zukin, *Landscapes of Power: From Detroit to Disney World* (Berkeley: University of California Press, 1991), 183.

17. Appropriately, the World Financial Center's Winter Garden Mall had been a setting for a scene in the film adaptation of Wolfe's story of New York's rich and powerful, *Bonfire of the Vanities*.

18. This respondent pointed out a rarely considered difficulty with modern urban life: with no running water, working on the thirty-seventh floor meant that finding a proper bathroom required walking down, and then up, thirty-seven flights of stairs. People did not always make that trip, so sanitary conditions were poor.

19. "F.B.I. Agents Raid Currency Offices." *New York Times*, November 19, 2003; Jonathan Fuerbringer and William K. Rashbaum, "Currency Fraud Ran Deep, Officials Say," *New York Times*, November 20, 2003.

20. On building Battery Park City, see Maynard T. Robison, "Rebuilding Lower Manhattan, 1955–1974" (PhD diss., City University of New York, 1976), and

Maynard T. Robison, "Vacant Ninety Acres, Well Located, River View," in *The Apple Sliced: Sociological Studies of New York City*, ed. Vernon Boggs, Gerald Handel, and Sylvia F. Fava (South Hadley, MA: Bergin and Garvey, 1984). On its development, see Fainstein, *City Builders*. For claims about the effect of the citadel, see Boyer, "Cities for Sale" and *City of Collective Memory*; David L. A. Gordon, *Battery Park City: Politics and Planning on the New York Waterfront* (Amsterdam: Gordon and Breach Press, 1997); Russell, "Battery Park City"; Sennett, *Conscience of the Eye*; Sorkin, *Variations on a Theme Park*.

21. Lopate, "Planners' Dilemma," quoted in Fainstein, *City Builders*, 183.

22. Boyer, *City of Collective Memory*, 450.

23. Russell, "Battery Park City," 208. Emphasis mine. On urban "authenticity," see Sharon Zukin, *Naked City: The Death and Life of Authentic Urban Places* (New York: Oxford University Press, 2010).

24. Thomas F. Gieryn, "A Space for Place in Sociology," *Annual Review of Sociology* 26 (2000): 463–96; Peter Dreier, John Mollenkopf, and Todd Swanstrom, *Place Matters: Metropolitics for the Twenty-First Century*, 2nd ed. (Lawrence: University Press of Kansas, 2005).

25. See, for instance, Fredric Jameson, *Postmodernism, or, The Cultural Logic of Late Capitalism* (Durham: Duke University Press. 1991), 42. As an example of reading the city like a text: "This diagnosis is confirmed by the great reflective glass skin of the Bonaventure [Hotel], whose function I will now interpret. . . . One would want . . . to stress the way in which the glass skin repels the city outside, a repulsion for which we have analogies in those reflector sunglasses which make it impossible for your interlocutor to see your own eyes and thereby achieve a certain aggressivity toward and power over the Other." I argue that whether a building plays an exclusive role can be determined by observing how people actually use it, or studying, historically, the role developers wanted it to play. Mirrored glass could have been used for too many reasons to be determinant on its own.

26. For an insightful account of nannies in affluent New York neighborhoods, see Tamara Mose Brown, *Raising Brooklyn: Nannies, Child Care, and Caribbeans Creating Community* (New York: NYU Press, 2011).

27. See Benjamin Heim Shepard and Gregory Smithsimon, *The Beach beneath the Streets: Contesting New York City's Public Spaces* (Albany: SUNY Press, forthcoming).

28. Richard Shepard, "Exploring Battery Park City," *New York Times*, May 19, 1989, C1.

29. Perhaps as a result of historic preservation movements, "history" is often conceived of as a charming asset to a neighborhood, decorating a place the way architectural detailing adorns Victorian homes. The fact that September 11 is now part of Battery Park City is a reminder that "history" carries much more weight than it is given in discussions that conceive of history as an asset to gentrifying neighborhoods.

30. Throughout the book, *neighborhood* refers to the physical place, and *community* describes the social group that occupies that space. The broad consensus among residents and others regarding the neighborhood's boundaries, and the spatial definition of community, led to greater congruence among these two concepts than usual.

31. The classic ethnography that includes domestic space is Carol Stack's *All Our Kin: Strategies for Survival in a Black Community* (New York: Basic Books, 1983); for a workplace ethnography, Rick Fantasia, *Cultures of Solidarity: Consciousness, Action, and Contemporary American Workers* (Berkeley: University of California Press, 1988); "Street corner" ethnographies include William F. Whyte, *Street Corner Society* (Chicago: University of Chicago Press, 1993); Elliott Liebow, *Tally's Corner* (Boston: Little, Brown, 1967).

32. Mitchell Duneier, *Sidewalk* (New York: Farrar, Straus and Giroux, 1999).

33. There is some evidence of a new interest in "studying up," or studying elites. See Miriam Greenberg, *Branding New York: How a City in Crisis Was Sold to the World* (New York: Routledge, 2008). Two recent ethnographies that include studies of elites (and workers) are Corey Dolgon's *The End of the Hamptons: Scenes from the Class Struggle in America's Paradise* (New York: NYU Press, 2005) and Rachel Sherman's study of workers and guests in luxury hotels, *Class Acts: Service and Inequality in Luxury Hotels* (Berkeley: University of California Press, 2007).

34. Even had rents not been at least double what I was able to pay, another concern was that contaminants like asbestos and heavy metals had inundated the buildings during the collapse of the Twin Towers. Environmental activists made a convincing case that the Environmental Protection Agency's cleanup had been inadequate. The risks, though uncertain, seemed too high, particularly for my two young children, Una and Eamon.

Notes to Chapter 1: Creating Battery Park City

1. For a fuller discussion of how large development projects plans provide some of the earliest indicators of elites' plans for the city, see Benjamin Heim Shepard and Gregory Smithsimon, *The Beach beneath the Streets: Contesting New York City's Public Spaces* (Albany: SUNY Press, forthcoming). The book also provides a longer description of privatized, filtered, community, and suburban spaces.

2. For more on defining public space, see Stephen Carr, Mark Francis, Leanne G. Rivlin, and Andrew M. Stone, *Public Space* (New York: Cambridge University Press, 1992); Lewis Dijkstra, "Public Spaces: A Comparative Discussion of the Criteria for Public Space," in *Constructions of Public Space: Research in Urban Sociology*, ed. Ray Hutchinson (Stamford, CT: JAI Press, 2000), 5:1–22.

3. As a working port, of course, the waterfront was closed off to the public. Raymond W. Gastil, *Beyond the Edge: New York's New Waterfront* (New York: Princeton Arch Press, 2002), 20.

4. Downtown-Lower Manhattan Association, *Major Improvements: Land Use, Transportation, Traffic* (New York: Downtown–Lower Manhattan Association, 1963), 12. See also Battery Park City Authority, "1962: Revitalization," www.battery parkcity.org/pages/page14.html.

5. Joshua Freeman, *Working-Class New York: Life and Labor since World War II* (New York: Free Press, 2000), 8. Most large cities had a higher *percentage* of workers in manufacturing than New York, but New York had larger raw numbers.

6. Sharon Zukin, *Naked City: The Death and Life of Authentic Urban Places* (New York: Oxford University Press, 2010), 223.

7. Robert Fitch, *The Assassination of New York* (New York: Verso, 1993).

8. New York City's three central business districts are Midtown, the Financial District, and Downtown Brooklyn. Plans in the last decades have envisioned a fourth near Long Island City, Queens, but beyond a single Citibank high-rise it has not taken shape. Across the Hudson River from Battery Park City is yet another central business district, Jersey City, whose office towers expanded dramatically over the course of this study.

9. Much of the history of Chase and Rockefeller's involvement comes from Maynard T. Robison, "Rebuilding Lower Manhattan, 1955–1974" (PhD diss., City University of New York, 1976). Eric Darton describes the Chase building as "David Rockefeller's opening gambit toward reshaping Lower Manhattan and leveraging the value of his family's real estate holdings there." Eric Darton, *Divided We Stand: A Biography of New York's World Trade Center* (New York: Basic Books, 1999), 13.

10. David Rockefeller, *Memoirs* (New York, Random House, 2002), 388.

11. Stephanie Strom, "Last of the Big-Time Rockefellers: Almost Alone, He Keeps the Flame Alive," *New York Times*, December 10, 1995.

12. Darton, *Divided We Stand*, 75.

13. Peter Collier and David Horowitz, *The Rockefellers: An American Dynasty* (New York: Holt, Rinehart and Winston, 1976).

14. Ibid., 415.

15. Robert Battaly, "Records of the Downtown-Lower Manhattan Association," Rockefeller Archive Center, January 2008, www.rockarch.org/collections/rockorgs/dlma.pdf.

16. "A World Center of Trade Mapped off Wall Street," *New York Times*, January 27, 1960, 1. At the beginning, a "world center of trade" was described as a complex housing offices, hotels, a securities exchange, and an exhibition hall for showcasing global products. See also "1,600-Unit Project for Battery Park Submitted to City," *New York Times*, June 10, 1961.

17. "Plans for Battery Park City are patterned after a waterfront development proposal prepared by the city's Department of Marine and Aviation in April, 1963. This was the first recent plan to call for the extension of Manhattan's shoreline through landfills." Henry Raymon, "Split on Planning Looms Downtown: State

Renewal Program for Area Conflicts with City's," *New York Times*, September 12, 1966, 38.

18. Anthony Flint, *Wrestling with Moses: How Jane Jacobs Took on New York's Master Builder and Transformed the American City* (New York: Random House, 2009), 148.

19. Robert A. Caro, *The Power Broker: Robert Moses and the Fall of New York* (New York: Alfred A. Knopf, 1974), 734. Quoted in Battaly, "Records."

20. Rockefeller, *Memoirs*.

21. Jonathan Rieder writes that to the middle- and working-class white Jewish and Italian residents he studied (and to the upper class as well), New York by the 1970s "had come to feel like an alien place. The perception of the physical environment as dangerous and unpredictable was grounded in a set of undeniable realities —the proximity of large numbers of lower-class ghetto dwellers, the eruption of conflicts in the 1960s between blacks and whites at all class levels, and the breakdown of old patterns of racial dominance and deference. These factors shaped the way Canarsians experienced material and symbolic space." *Canarsie: The Jews and Italians of Brooklyn against Liberalism* (Cambridge, MA: Harvard University Press, 1985), 57.

22. Derek Edgell, *The Movement for Community Control of New York City's Schools, 1966–1970: Class Wars* (Lewistown, NY: Edwin Mellen Press, 1988), 12.

23. Eric M. Javits, *SOS New York: A City in Distress* (New York: Dial Press, 1961), 128.

24. Though born and raised in New York, Whalen had moved to Long Island. Richard J. Whalen, *A City Destroying Itself: An Angry View of New York* (New York: William Morrow, 1965), 7.

25. Ibid., 18.

26. Edgell, *Movement for Community Control*, 12.

27. Murray Illson, "Fear Said to Keep City Parks Empty: Civic Group Says Its Survey Shows People Are Afraid Even during Daylight," *New York Times*, May 20, 1963, 33.

28. As Joe Flood points out, narratives of cities and safety are often selectively constructed out of the countless potential data points that could construct a narrative of safety or disorder. In this case, rising arrests could be a sign of increased enforcement, not increased crime. And Jane Jacobs would suggest that parks are never safe places to be at night because there are few "eyes on the street." Both facts, however, were used at the time as evidence the parks were descending into chaos. On how a narrative of "city in crisis" makes sense of "out-of-context crime stats," see Joe Flood, *The Fires: How a Computer Formula, Big Ideas, and the Best of Intentions Burned Down New York City—and Determined the Future of Cities* (New York: Riverhead Books, 2010), 114–15.

29. Murray Illson, "Parks Fears Scored as Exaggerated: Police Aide Says Im-

pression of Public Is Erroneous," *New York Times*, September 28, 1963, 21, 44; italics mine.

30. Battery Park City Authority, "Master Plan: Integrated Society," 1966, www .batteryparkcity.org/page/page14_2.html.

31. Wallace K. Harrison, *"Battery Park City": New Living Space for New York, a Proposal for Creating a Site for Residential and Business Facilities in Lower Manhattan, 1966*, Wallace K. Harrison Archives, Avery Library, Columbia University (handwritten on cover: "Battery Park My Start"), (New York: Harrison and Abramovitz, Architects, 1966), 13.

32. Ibid.

33. Gregory Smithsimon, "Dispersing the Crowd: Bonus Plazas and the Creation of Public Space," *Urban Affairs Review* 43 no. 3 (January 2008): 325–51.

34. *Growth machine* is Logan and Molotch's term for the coalition of developers, government, corporations, and city institutions that support large-scale urban development. See further discussion below. John Logan and Harvey Molotch, *Urban Fortunes: The Political Economy of Place* (Berkeley: University of California Press, 1988).

35. Carol Willis, ed., *The Lower Manhattan Plan: The 1966 Vision for Downtown New York* (New York: Princeton Architectural Press, 2002).

36. Deirdre Carmody, "Cheering Throng Jams Fulton Street for Opening," *New York Times*, July 29, 1983.

37. Joshua Olsen, "James Rouse Made a Difference in the American Landscape," n.d., Columbia Archives, www.columbiaarchives.org/?action=content.sub &page=biography_1&oid=1.

38. Among other things, Rouse built the planned, racially integrated community of Columbia, Maryland, and advised on some of the earliest federal urban renewal programs. See Nicholas Dagen Bloom, *Merchant of Illusion: James Rouse: America's Salesman of the Businessman's Utopia* (Ohio State University Press, 2004); Joseph Rocco Mitchell and David L. Stebenne, *New City upon a Hill: A History of Columbia, Maryland* (Charleston, SC: History Press, 2007).

39. New York City Planning Commission, *Plan for New York City: A Proposal* (New York: Department of City Planning, 1969).

40. "Affront to the needy" from Maynard T. Robison, "Vacant Ninety Acres, Well Located, River View," in *The Apple Sliced: Sociological Studies of New York City*, ed. Vernon Boggs, Gerald Handel, and Sylvia F. Fava (South Hadley, MA: Bergin and Garvey, 1984).

41. Robison, "Rebuilding Lower Manhattan"; see also Robison, "Vacant Ninety Acres," 192.

42. Robison, "Vacant Ninety Acres."

43. Ibid., 186.

44. Ibid., 188.

45. Darton, *Divided We Stand*. Minoru Yamasaki, the Trade Center's architect,

was also the architect of the Pruitt-Igoe Housing Project in St. Louis, which was greeted as a modernist breakthrough at its dedication in 1955 but was derided as a failure when the city demolished it by controlled explosion seventeen years later (118–20). Observers have suggested that the Port Authority chose Yamasaki to design the original tower, and Daniel Libeskind to design its replacement, because both architects' accommodating dispositions ensured that the Port Authority would be able to achieve its ambitious programmatic goals without the interference that other high-profile designers might cause. Yamasaki would not let stylistic objectives interfere with the goal of maximizing square footage. Libeskind's personal presentations in Lower Manhattan certainly suggested he sought to go along with whatever his client wanted, even if it meant making excuses for changes to core elements of his own plan. See Robin Pogrebin, "The Incredible Shrinking Daniel Libeskind," *New York Times*, June 20, 2004.

46. Gregory Smithsimon, "The Technologies of Public Space and Alternatives to a Privatized New York," paper presented at the annual meetings of the American Sociological Association, August 2001.

47. Oscar Newman, *Defensible Space: Crime Prevention through Urban Design* (New York: Macmillan, 1972), 3.

48. Ibid., 18.

49. Ibid., 3; italics mine.

50. Ibid., 1.

51. Whalen nodded approvingly at the comments of an angry New Yorker who wrote, resentfully, after a civil rights march in Alabama, "We also need a great civil rights march in our city to insure to us the civil rights to live in our homes, to ride in our subways, to walk in our streets and parks at any hour without fear of being murdered, robbed and raped." Whalen, *City Destroying Itself*, 23–24.

52. Mayor Abraham Beame, "Battery Park City Is Alive and Well and Growing in NYC!" ca. 1976, Mayor Abraham Beame Papers, New York Municipal Archives.

53. Battery Park City Authority, "1975—Defensible Space," under "Who We Are," "Urban Experiment," n.d., www.batteryparkcity.org/page/page14_4.html, accessed May 13, 2010.

54. David L. A. Gordon, *Battery Park City: Politics and Planning on the New York Waterfront* (Amsterdam: Gordon and Breach, 1997), 50.

55. Kirk Johnson, "A Green Foothold in the Downtown Concrete," *New York Times*, October 15, 2002, B1.

56. Jack Byers, "The Privatization of Downtown Public Space: The Emerging Grade-Separated City in North America," *Journal of Planning Education and Research* 17 (Spring 1998): 189–205.

57. The North neighborhood buildings in this plan were to be built with mortgages subsidized under Section 236 of the Housing Act of 1968. Battery Park City, "Master Plan: 1975—Defensible Space," www.batteryparkcity.org/page/page14_4 .html; Beame, "Battery Park City Is Alive."

58. Robert D. McFadden, "Abraham Beame Is Dead at 94; Mayor during 70's Fiscal Crisis," *New York Times*, February 11, 2001.

59. At an honorary lunch years later, then-mayor Abe Beame was given a life preserver for having "saved" New York. Beame tried to dispute the analogy, saying, "The city was never in the position of the Titanic." Others disagreed. See Joyce Purnick, "Beame's Goal: Setting History Straight," *New York Times*, August 3, 1995.

60. Citibank's move ultimately put the bankers, as creditors, in charge of New York City's finances. Through the Municipal Assistance Corporation that oversaw New York's budget in this period, financial elites ensured that regular New Yorkers would suffer rather than banks, by slashing civil servant wage costs and funding for social welfare programs and then funneling the savings into banks' coffers. On Walter Wriston's role, see David Harvey, *A Brief History of Neoliberalism* (New York: Oxford University Press, 2007), 45. Other sources reference the president of Citibank, William Spencer, in a less singular role; see Joshua Freeman, *Working-Class New York*, 257.

61. In an interview with me, the architect of record for the Citibank Tower, Richard Roth of Emery Roth and Sons, explained that the slanted building top had been an aesthetic decision, not an environmental one. Initially, in fact, it was drawn up so that the slanted side would face northward. Someone suggested the angled roof could support solar energy panels, and the designed was simply flipped so that it had a sunnier southern exposure. There had indeed been plans to put solar water heating equipment on the south-facing roof, but when designers calculated that the solar technology wouldn't pay back the initial investment for a hundred years, the plan was unceremoniously scrapped. The roof, however, still faces the sun.

62. Logan and Molotch, *Urban Fortunes*.

63. David Firestone, "3 Tell Council They Beat Homeless to Clear Out Business District," *New York Times*, May 11, 1995; Thomas J. Lueck, "Grand Central Partnership Is Subject of U.S. Inquiry," *New York Times*, May 26, 1995. For a revealing history of the use of Penn Station by the homeless, see Mitchell Duneier, *Sidewalk* (New York: Farrar Straus, and Giroux, 2000).

64. Clifford Krauss, "Special Unit Ushers Homeless from Subways," *New York Times*, September 4, 1994.

65. Bruce Lambert, "Neighborhood Report: Union Square; Confronted by the Homeless Domino Effect, Another Park Cracks Down," *New York Times*, June 12, 1994. For the definitive account of the sweeping of Tompkins Square Park, see Janet Abu-Lughod, ed., *From Urban Village to East Village: The Battle for New York's Lower East Side* (New York: Wiley-Blackwell, 1995).

66. Rick Bragg, "Police Captain Apologizes for Remarks on Homeless," *New York Times*, July 20, 1994.

67. "Homeless Plan Needs More Work" (editorial), *New York Times*, May 17, 1994.

68. Shawn G. Kennedy, "Subsidy Cuts Raise Concern for Homeless," *New York Times*, September 11, 1994.

69. Raymond Hernandez, "Suit Faults New York City over Upstate Homeless Site," *New York Times*, November 23, 1994.

70. For examples of state subsidies of suburban development, from federal highway funds to, most significantly, mortgage interest tax deductions, see Kenneth T. Jackson, *Crabgrass Frontier: The Suburbanization of the United States* (New York: Oxford University Press, 1985), 293–95.

71. Zukin, *Naked City*, 222–23.

72. Alexander Cooper Associates, *Battery Park City Draft Summary Report and 1979 Master Plan*, 1979, www.batteryparkcity.org/pdf_n/1979_Master_Plan.pdf, 42; italics mine.

73. Ibid.; italics mine.

74. Among other references to the imperial quality of global cities, see Anthony D. King, *Global Cities: Post Imperialism and the Internationalization of London* (New York: Routledge, 1990).

75. Jane Jacobs, *The Death and Life of Great American Cities* (New York: Vintage, 1961).

76. Flint, *Wrestling with Moses*.

77. "Top Projects Started, 2005–2006," New York Construction, June 2006, www.newyork.construction.com/projects/TopPrj_05-06/TPstrt1-5.pdf.

78. Other activities came to light in addition to Goldman's involvement in speculation in the housing market bubble and its efforts to conceal the true extent of the Greek government's financial difficulties. Gretchen Morgenson and Louise Story, "Banks Bundled Bad Debt, Bet against It and Won," *New York Times*, December 23, 2009. In April 2010, one Goldman trader faced securities fraud charges related to selling investments he had designed to fail (and then betting against the investments), and CEO Lloyd C. Blankfein testified before Congress, denying that Goldman had systematically bet against American homeowners. "Testimony from Lloyd C. Blankfein Chairman and CEO, the Goldman Sachs Group, Inc., Permanent Senate Subcommittee on Investigations, April 27, 2010," www.google.com/search?client=safari&rls=en&q="Testimony+from+Lloyd+C.+Blankfein+Chairman+and+CEO%22&ie=UTF-8&oe=UTF-8. See also Christine Hauser, "Investors Were Not Duped, Goldman Tells Senators," *New York Times*, April 27, 2010.

79. Calvin Tomkins, "Big Art, Big Money: Julie Mehretu's 'Mural' for Goldman Sachs," *New Yorker*, March 29, 2010, 62–69.

Notes to Chapter 2: Real Privilege and False Charity

1. Battery Park City held 4,400 households, and New Settlement had 995 families. New Settlement Apartments, "Annual Report of New Settlement Apart-

ments," March 2004, author's private collection. Data from Battery Park City come from the 2000 U.S. Census.

2. Paulo Freire, *Pedagogy of the Oppressed* (New York: Seabury Press, 1970).

3. James C. McKinley Jr., "Spitzer and Hevesi Propose Overhauling Operation of State Authorities," *New York Times*, February 25, 2004, B6.

4. City Project, "The Failed Promise of Battery Park City: Housing and Governance Issues along the Hudson," September 2000, author's private collection.

5. Ibid.

6. David L. A. Gordon, *Battery Park City: Politics and Planning on the New York Waterfront* (Amsterdam: Gordon and Breach Press, 1997).

7. Maynard T. Robison, "Vacant Ninety Acres, Well Located, River View," in *The Apple Sliced: Sociological Studies of New York City*, ed. Vernon Boggs, Gerald Handel, and Sylvia F. Fava (South Hadley, MA: Bergin and Garvey, 1984), 187.

8. Ibid., 186.

9. Gordon, *Battery Park City*, 101.

10. Eric Lipton, "Missing Element/A Special Report; Battery Park City Is Success, Except for Pledge to the Poor," *New York Times*, January 2, 2001.

11. Douglas S. Massey and Nancy A. Denton, *American Apartheid: Segregation and the Making of the Underclass* (Cambridge, MA: Harvard University Press, 1993); Eduardo Bonilla-Silva, *Racism without Racists: Color-Blind Racism and the Persistence of Racial Inequality in the United States*, 2nd ed. (New York: Rowman and Littlefield, 2006).

12. Bonilla-Silva, *Racism without Racists*.

13. The description of New York liberals as the "best and the brightest" comes from Joe Flood, *The Fires: How a Computer Formula, Big Ideas, and the Best of Intentions Burned Down New York City—and Determined the Future of Cities* (New York: Riverhead Books, 2010). See also Larry Williams's obituary for David Halberstam, "Journalist Chronicled the Culture of America," *Baltimore Sun*, April 24, 2007.

14. For a searing account of the planned shrinkage policy, see Deborah and Roderick Wallace, *A Plague on Your Houses: How New York Was Burned Down and National Public Health Crumbled* (New York: Verso, 1999).

15. No Battery Park City residents volunteered an explicit preference for a white neighborhood. Two things reduced the chances of my hearing this kind of comment: first, because I was a graduate student researcher, residents probably would not have thought my political views would make me welcome such comments. Second, because my research put me in touch disproportionately with neighborhood activists, I interacted with a more "civic-minded" subset of the community, one that was less likely to express such views. Residents did laud the privacy Battery Park City afforded but did not cast it in racial terms, and rarely even in terms that are often code for racial preferences.

16. On color-blind ways people talk about race without talking explicitly about

race, see Bonilla Silva, *Racism without Racists*. For research on whites' preferences for predominantly white neighborhoods, see Bonilla Silva as well as Massey and Denton, *American Apartheid*.

17. Massey and Denton, *American Apartheid*, 230.

18. "Lower Manhattan Development Corporation Approves Revised Version of Individual Assistance Plan," press release, April 9, 2002, Lower Manhattan Development Corporation, www.renewnyc.com/displaynews.aspx?newsid=87609911-c567-4755-8fa5-638335ac3104.

19. Ida Susser, *Norman Street: Poverty and Politics in an Urban Neighborhood* (New York: Oxford University Press, 1982).

20. Susan J. Fainstein, *The City Builders: Property, Politics, and Planning in London and New York* (Cambridge, MA: Blackwell, 1994), 266 n. 35.

21. Shawn G. Kennedy, "Subsidy Cuts Raise Concern for Homeless," *New York Times*, September 11, 1994.

Notes to Chapter 3: Residents, Space, and Exclusivity

1. David W. Dunlap, "At Battery Park City, 'Pioneers' Like Life," *New York Times*, July 22, 1983, B1.

2. Jerry Cheslow, "If You're Thinking of Living in Battery Park City; A New Neighborhood along the Hudson," *New York Times*, December 26, 1993, 10:3.

3. Alex Krieger, "Reinventing Public Space," *Architectural Record* 183 (June 1995): 76–77.

4. Iris Marion Young, "The Ideal of Community and the Politics of Difference," in *Feminism/Postmodernism*, ed. Linda J. Nicholson (New York: Routledge, 1990), 300–23.

5. Richard Sennett, *The Fall of Public Man* (New York: Norton, 1976), 301–11.

6. Peter Marcuse, "The Enclave, the Citadel, and the Ghetto: What Has Changed in the Post-Fordist U.S. City," *Urban Affairs Review* 33, no. 2 (1997): 249.

7. Even in communities far less privileged than Battery Park City, community has been the basis of violently exclusive organizing—for instance, that of working-class white communities to exclude African Americans. See Thomas Sugrue, *The Origins of the Urban Crisis: Race and Inequality in Postwar Detroit* (Princeton: Princeton University Press, 1996).

8. George W. Goodman, "At Battery Park City, a Rent Strike," *New York Times*, January 16, 1983, 8:6; Dunlap, "At Battery Park City," B1.

9. Susan Chira, "Battery Park City Rousing from a 2-Year Fiscal Sleep," *New York Times*, November 7, 1989, B1.

10. Peter Malbin, "If You're Thinking of Living in Battery Park City; Urban Suburb w/Yacht Basin," *New York Times*, May 19, 1998, 11:3.

11. David W. Dunlap, "Opening New Fronts at Battery Park City," *New York Times*, September 4, 1994.

12. Sarah Lyall, "Battery Park City Moves to Adolescence," *New York Times*, June 30, 1987, B1.

13. "New South Park Design Gets Warmer Reception," *Downtown Express*, March 8, 1994, 4, quoted in David L. A. Gordon, *Battery Park City: Politics and Planning on the New York Waterfront* (Amsterdam: Gordon and Breach, 1997), 95.

14. For more on collective memory, see Jeffrey K. Olick, *In the House of the Hangman: The Agonies of German Defeat, 1943–1949* (Chicago: University of Chicago Press, 2005).

15. Dunlap, "At Battery Park City."

16. For an example of threats to class status, see the chapter "Homegrown Revolution," in Mike Davis, *City of Quartz: Excavating the Future in Los Angeles* (New York: Verso, 1990), 151–263. See also Andrew Weise, *Places of Their Own: African American Suburbanization in the Twentieth Century* (Chicago: University of Chicago Press, 2004), 282–83.

17. Alexander Cooper Associates, *Battery Park City Draft Summary Report and 1979 Master Plan*, 1979, www.batteryparkcity.org/pdf_n/1979_Master_Plan .pdf, 62.

18. Steven Flusty, *Building Paranoia: The Proliferation of Interdictory Space and the Erosion of Spatial Justice* (Los Angeles: Los Angeles Forum for Architecture and Urban Design, 1994), 17.

19. Julia Trilling, "A Future That Looks Like the Past: Planners in Downtown San Francisco and at Battery Park City, in New York, Are Trying to Design New Buildings That Look Like They Belong Next to Old Ones," *Atlantic*, July 1985, 28–34.

20. Battery Park City Authority, "Battery Place Residential Area: Design Guidelines, May 1985, amended February 1989," p. 4, Avery Library, Columbia University.

21. Battery Park City Authority, *Battery Park City Design Guidelines for the North Residential Neighborhood* (New York: Battery Park City Authority, 1994), 15. Repeated in design guidelines for other sections.

22. Setha Low, *Behind the Gates: Life, Security, and the Pursuit of Happiness in Fortress America* (New York: Routledge, 2003).

23. Jeffrey K. Olick, *States of Memory: Continuities, Conflicts, and Transformations in National Retrospection*, ed. Jeffrey K. Olick (Durham: Duke University Press, 2003), 8.

24. Phillip Lopate, *Waterfront: A Journey around Manhattan* (1989; repr., New York: Crown, 2004), 28.

25. Manuel Castells, *The City and the Grassroots: A Cross-cultural Theory of Urban Social Movements* (Berkeley: University of California Press, 1983), 311.

26. I compared 2000 census figures for the population of Battery Park City and New York City to park acreage measured by Battery Park City Authority and Battery Park City Parks Conservancy (thirty-two acres), and for New York overall

(twenty-eight thousand acres) by the Trust for Public Land, in its report "NYC Ranks High in Park Acres, Low in Funding," December 3, 2001, www.tpl.org/tier3_cd.cfm?content_item_id=6120&folder_id=631.

27. For descriptions of other upper-class communities, see Low, *Behind the Gates*, and Corey Dolgon, *End of the Hamptons: Scenes from the Class Struggle in America's Paradise* (New York: NYU Press, 2005).

28. William Foote Whyte, *Street Corner Society: The Social Structure of an Italian Slum* (Chicago: University of Chicago Press, 1943); Elijah Anderson, *Streetwise: Race, Class and Change in an Urban Community* (Chicago: University of Chicago Press, 1990).

29. Steven Gregory, *Black Corona: Race and the Politics of Place in an Urban Community* (Princeton: Princeton University Press, 1998), 250.

30. Ibid., 149.

31. Sudhir Alladi Venkatesh, *American Project: The Rise and Fall of a Modern Ghetto* (Cambridge, MA: Harvard University Press, 2000), 3.

32. Jonathan Rieder, *Canarsie: The Jews and Italians of Brooklyn against Liberalism* (Cambridge, MA: Harvard University Press, 1985), 132–67.

33. Eight Battery Park City residents died in the World Trade Center, according to "September 11, 2002," *Battery Park City Broadsheet*, September 21–October 6, 2002, 2.

34. To a slight degree tenure appears to have helped too, since some longer-term residents expressed a disconnection between pre–September 11 and "new" residents. But many denied this was true, others were careful to stress that they welcomed the arrival of new residents, and even those who believed there was a difference hesitated to articulate it.

35. An insightful study on the role of kin is Carol Stack, *All Our Kin: Strategies for Survival in a Black Community* (New York: Basic Books, 1983). Other studies of poor and working-class families have demonstrated similar reliance on kin as a basic support network; see Kathryn Edin and Maria Kefalas, *Promises I Can Keep: Why Poor Women Put Motherhood before Marriage* (Berkeley: University of California Press, 2005).

Notes to Chapter 4: Oasis to Epicenter

1. "Eleanor Rosen" is a pseudonym.

2. The recovery of remains in the Winter Garden would seem initially to contradict a statement by Battery Park City Authority officials, quoted in the *Broadsheet* article reproduced in Appendix A, that no one was killed in Battery Park City. The residents saw the informal ceremony being conducted by recovery workers as remains of firefighters were removed from the Winter Garden. The firefighters had evidently been in the World Trade Center when the building collapsed into the

Winter Garden. In a disaster of this scope, there is uncertainty about where people died, as in the case of Battery Park City resident Sneha Anne Philip, who was last seen September 10 and is presumed to have perished in the attacks.

3. "Linda Edwards" is a pseudonym.

4. On the differential access to environmental information, see Jed Tucker, "Making Difference in the Aftermath of the September 11th 2001 Terrorist Attacks," *Critique of Anthropology* 24, no. 1 (2004): 34–50.

5. "Winter Garden Opens," *Battery Park City Broadsheet*, October 6, 2002, 3.

6. Kathleen J. Tierney, "Strength of a City: A Disaster Research Perspective on the World Trade Center Attack," 2001, Social Science Research Council, After September 11 Archive, www.ssrc.org/sept11/essays/tierney.htm.

7. Fran Tonkiss, *Space, the City and Social Theory* (Malden, MA: Polity Press, 2005), 72. See also Lewis Dijkstra, "Public Spaces: A Comparative Discussion of the Criteria for Public Space," in *Constructions of Public Space: Research in Urban Sociology*, ed. Ray Hutchinson (Stamford, CT: JAI Press, 2000), 5:1–22.

8. For more discussion of community space, see Benjamin Heim Shepard and Gregory Smithsimon, *The Beach beneath the Streets: Contesting New York City's Public Spaces* (Albany: SUNY Press, forthcoming).

9. Jane Jacobs's account of public space, in contradicting prevailing views at the time of cities as chaotic places, presents instead a liberal model of public space as a site of social harmony, irrespective of material and power differences among users; see her *Death and Life of Great American Cities* (New York: Vintage, 1961). Portrayals of public spaces as sites of recreation, consumption, and leisure are well represented in the planning literature. See also Ray Oldenburg, *The Great Good Place: Cafés, Coffee Shops, Community Centers, Beauty Parlors, General Stores, Bars, Hangouts and How They Get You through the Day* (New York: Paragon House, 1989). Idealizations of public spaces as the embodiment of a supposedly democratic public sphere—"the Greek agora, the coffeehouses of early modern Paris and London, the Italian piazza, the town square"—are discussed in Margaret Crawford, "Blurring the Boundaries: Public Space and Private Life," in *Everyday Urbanism*, ed. John Chase, Margaret Crawford, and John Kaliski (New York: Monacelli Press, 1999), 23.

10. The necessity that good public space allow users to control the degree of "entanglements" with others they meet there (thus encouraging contact by limiting obligations) is discussed further in Jacobs, *Death and Life*, 63–64.

11. Kai Erikson, *A New Species of Trouble: The Human Experience of Modern Disasters* (New York: Norton, 1994), 235–36; also C. E. Fritz, "Disasters," in *Social Problems*, ed. Robert Merton and R. Nisbet (New York; Harcourt, Brace and World, 1961), 651–94, and A. H. Barton, *Communities in Disaster* (Garden City, NY: Anchor, Doubleday, 1970), both quoted in G. A. Kreps, "Sociological Inquiry and Disaster Research," *Annual Review of Sociology* 10 (1984): 309–30. See also William Graham Sumner on comradeship in the "we-group" during times of war,

quoted in Robert D. Putnam, *Bowling Alone: The Collapse and Revival of American Community* (New York: Simon and Schuster), 267.

12. Erikson, *New Species of Trouble*, 235, 236; italics in original.

13. For other disasters where survivors recovered in isolation, see Peter E. Hodgkinson and Michael Stewart, *Coping with Catastrophe: A Handbook of Post-disaster Psychosocial Aftercare*, 2nd ed. (New York: Routledge, 1998).

14. Kai Erikson, *Everything in Its Path: Destruction of Community in the Buffalo Creek Flood* (New York: Simon and Schuster, 1978), 47.

15. Douglas M. Glandon, Jocelyn Muller, and Astier M. Almedom, "Resilience in Post-Katrina New Orleans, Louisiana: A Preliminary Study," *African Health Sciences* 8, no. 1 (2008): S21–S27. See also Chris Kromm and Sue Sturgis, "Hurricane Katrina and the Guiding Principles on Internal Displacement: A Global Human Rights Perspective on a National Disaster," Institute for Southern Studies Special Report, *Southern Exposure* 36, nos. 1–2 (January 2008), www.southernstudies.org/ISSKatrinaHumanRightsJan08.pdf.

16. David Abramson, Tasha Stehling-Ariza, Richard Garfield, and Irwin Redlener, "Prevalence and Predictors of Mental Health Distress Post-Katrina: Findings from the Gulf Coast Child and Family Health Study," *Disaster Medicine and Public Health Preparedness* 2, no. 2 (2008): 77–86.

17. Erikson, *New Species of Trouble*, 19.

18. Erikson focused on human-caused disasters in part because he conducted research in order to testify on behalf of plaintiffs against those who caused the disasters.

19. Erikson, *New Species of Trouble*, 239.

20. National Commission on Terrorist Attacks upon the United States, *The 9/11 Report* (New York: St. Martin's Press, 2004).

21. "Social as well as physical dimensions": see Kreps, "Sociological Inquiry"; Charman Pincha, "Indian Ocean Tsunami through the Gender Lens: Insights from Tamil Nadu, India," Oxfam, 2008, www.gdnonline.org/resources/Pincha_IndianOceanTsunamiThroughtheGender%20Lens.pdf.

Notes to Chapter 5: Every Day Is September 11

1. Because there has been periodic contention in the city between cyclists and pedestrians over biking on sidewalks, it is worth pointing out that cycling was allowed on the promenade.

2. Two ideas that had gained grassroots support outside Battery Park City were roundly dismissed by Battery Park City residents. The first, rebuilding the original Twin Towers, horrified residents, some of whom had lived through their being bombed twice, in 2001 and in 1993. They could not imagine having to look at, live near, or worry about identical towers once again looming outside their windows. The second suggestion, building nothing on the entire Trade Center site

(Mayor Rudolph Giuliani at one point proposed the whole area be a parklike memorial) was also unpopular; residents saw no benefit to making permanent the physical destruction of the area across the street from them.

3. The Authority minimized the memorial in several ways. They held the memorial competition last, after space had been assigned to all of the other uses (like commercial office buildings, retail, and transportation). Second, the memorial space was about the smallest it could be: it did not seem publicly acceptable for the Authority to build office towers or a shopping mall on the actual footprints of the towers, so the memorial was limited to those footprints. Finally, after the winning memorial design proposed that the memorial go "down to the bedrock," the Authority instead built train tracks and an electrical substation on the bedrock of the footprints. It seems sufficient to attribute their actions to money-making motives (even if they are a not-for-profit state agency). But the Authority often seemed to be actively trying to minimize the size and impact of the memorial as an end in itself. The Authority's actions raise the question of whether a bureaucracy can suffer survivor guilt and try to obscure a painful past.

4. Laura Davis-Chanin to *Battery Park City Broadsheet*, October 21, 2002, 4.

5. There is another commonality to this generation of memorials, which commemorated 58,195 Americans killed in the Vietnam War, 168 lives lost in the 1995 Oklahoma City bombing, and the nearly three thousand killed in the 2001 and 1993 bombings of the World Trade Center. Beginning with Maya Lin's monument, each has created a visceral sense of awe through the visual presentation of large numbers. As I experienced the Vietnam memorial, a visitor is slowly inundated in a rising tide of names until their number is overwhelming, just as the war itself progressed first as a trickle and then as a flood. In Oklahoma, visitors are faced with an array of 168 empty seats, one for each person killed. A similar identification of individuals by name is planned for the Trade Center memorial. Size and scale have long been used to create a humbling sense of awe. In monuments constructed before the era of skyscrapers, physical scale—massive arches or towering obelisks —was used to create a sense of the sublime. In the contemporary examples, the awe triggered in human cognition when presented by large numbers is transmogrified into a humbling awe at the scale of loss of life.

I speculate that these monuments create a sense of awe because of their visual presentation of quantities beyond the capacity of humans to easily comprehend. Such monuments present a number of people above "Dunbar's number" (often given as 150 people), said by Gladwell to be the "cognitive limit to the number of individuals with whom any one person can maintain stable relationships." The anthropologist Robin Dunbar speculates that humans have cognitive difficulty keeping track of more than this number of individuals and finds that human social groupings (such as villages and nomadic tribes) rarely exceed this number. A memorial that presents a viewer with a large number of people creates a disorienting sense of wonder perhaps not unlike that perceived by a twelfth-century person looking at a very tall cathedral. A different approach would be required for a moving monument to a smaller number of deaths. For more on the calculation of Dunbar's number (including calculations based on "cranial volume," which should always give social scientists pause), see Robin Dunbar, "Neocortex Size as a Constraint on Group Size in Primates," *Journal of Human Evolution* 22, no. 6 (1992): 469–93. The most successful popularization of Dunbar's number is Malcolm Gladwell, *The Tipping Point: How Little Things Can Make a Big Difference* (New York: Back Bay Books, 2002).

6. From photos and online satellite maps, the memorial to the Oklahoma City bombing fits with the trend of most memorials; homes are far enough away that few people will cross the memorial without intending to visit it. If New Orleans ever establishes a substantial commemoration to Hurricane Katrina, it will probably have to address questions of location like those faced in Lower Manhattan.

7. Jeff Goldman is a pseudonym.

8. "September 11, 2002," *Battery Park City Broadsheet*, September 21–October 6, 2009, 2.

9. William H. Whyte, *City: Rediscovering the Center* (New York: Doubleday, 1988).

10. Ibid.

Notes to Chapter 6: Class and Community Organizations

1. Contrary to the popular story that has grown up around Fitzgerald's line, Hemingway never replied, "Yes. They have more money."

2. F. Scott Fitzgerald, "The Rich Boy," 1926, published in the public domain by Feedbooks (www.feedbooks.com/book/1153).

3. The managers of the World Financial Center have been significant players because they control the tallest buildings and most central location in the neighborhood. Olympia and York, which developed and ran the property, went bankrupt. Today, Brookfield Properties runs the World Financial Center. Like the grassroots organizations, the corporate entities in this concentrated community showed a significant degree of interconnection. Because the land is owned by the Battery Park City Authority, the Lefrak, Rockrose, and Brookfield management companies are tenants of Battery Park City Authority, holding long-term leases. Meyer "Sandy" Frucher, former president of the Battery Park City Authority, later worked for Olympia and York. While there is no evidence of particular connections among the management companies, the Authority's regular dealings with each of them facilitated work between the Authority and any one company on issues outside their tenancy, such as development, use, and renovation of surrounding spaces.

4. So many groups were founded by Gateway residents that they could not all be discussed here. A sampling of those *not* discussed in the book are the

Downtown Community Restoration Project (Tammy Meltzer), the BPC Dog Association (Jeff Galloway), and Stockings with Care (Rosalie Joseph). For profiles of each, see "Everyday Heroes," *Battery Park City Broadsheet*, February 12–27, 2002.

5. The 1995 Oklahoma City bombing has often been called the worst terrorist attack in American history before September 11. Without in anyway diminishing the horror of the deaths of 168 people in that attack, in fact, it was not even the worst terrorist attack in Oklahoma history. The Tulsa Race Riots of 1921 killed three hundred people and left ten thousand homeless. For that reason, I hesitate to assign even an attack as gruesome as the September 11 attacks the title of the worst in U.S. history. A ten-minute walk from the Trade Center site, one can see where the British prison ships held U.S. prisoners of war and killed over eleven thousand. Across the Hudson River is the site of the Black Tom Island explosion, where in 1916 foreign agents blew up a munitions depot, causing the equivalent of a 5.0 magnitude earthquake. A quick train ride brings one to the Bronx or Brooklyn, where an estimated three thousand people were killed in building fires spurred on by an official city policy of neglect of poor black and Latino neighborhoods in the 1960s, '70s, and '80s. The September 11 attacks were without precedent in my life's experience. But even local history, unfortunately, makes superlatives difficult to assign.

6. See, for instance, Herbert Gans's discussions of cosmopolitan versus conservative segments of the middle class in *The Levittowners: Ways of Life and Politics in a New Suburban Community* (New York: Columbia University Press, 1967).

7. Both a survey by Friends of Community Board 1 in the spring of 2003 and another by Pace University in March 2004 found widespread opposition to the West Street tunnel. In such surveys, residents generally worried, as grassroots groups did, about a vibrant, mixed-use design for the Trade Center site and less about the memorial itself.

8. Dave Stanke, "Limit the Remains at Ground Zero," *Daily News*, March 14, 2003, 35.

9. "Celebrate the Spirit of Battery Park City at the First Annual Block Party," *Battery Park City Broadsheet*, October 6, 2002, 1.

10. CERT stands for Community Emergency Response Team.

11. Richard Perez-Pena, "Downtown, State Plans Rebuilding Agency, Perhaps Led by Giuliani," *New York Times*, November 3, 2001, B10.

12. Campaign contribution figures come from Federal Election Commission records compiled by the Center for Responsive Politics' Donor Lookup database at opensecrets.org. Whitehead's former position and political connections are described in Robin Finn, "Public Lives: The Reluctant Director of Downtown Restoration," *New York Times*, January 18, 2002, B2.

13. Frank Lombardi, "Rudy, Gov. Ready to Rebuild: Unveil Corp. to Redevelop Lower Manhattan," *Daily News* (New York), November 3, 2001, 15.

14. On advisory boards, see Pete Donohue and Eric Herman, "WTC Advisory

Group Set: Sacred Ground Status Pressed for Twin Sites," *Daily News* (New York), January 26, 2002, 17.

15. Lower Manhattan Development Corporation, "World Trade Center Memorial and Redevelopment Plan Final Generic Environmental Impact Statement," April 2004, www.renewnyc.com/plan_des_dev/environmental_impact_contents_april2004.asp, quoted in e-mail message from Coalition to Save West Street, "4/16/04 West St. Advisory: Battery Park City Bus Garage Gone; State DOT Quash," April 17, 2004.

16. Coalition to Save West Street, e-mail, "4/16/04 West St. Advisory: Battery Park City Bus Garage Gone; State DOT Quash," April 17, 2004.

17. Coalition to Save West Street, e-mail, "4/24/04 West St. Advisory: Schumer Opposes Tunnel; State DOT Apologizes!" April 24, 2004.

18. Herbert J. Gans, *The Urban Villagers: Group and Class in the Life of Italian-Americans* (New York: Free Press, 1962), 8.

19. Sudhir Alladi Venkatesh, *American Project: The Rise and Fall of a Modern Ghetto* (Cambridge, MA: Harvard University Press, 2000); Steven Gregory, *Black Corona: Race and the Politics of Place in an Urban Community* (Princeton: Princeton University Press, 1998).

20. In fact, they formed the Save the West End Committee, a name that echoed in the Battery Park City group the Coalition to Save West Street.

21. Gans, *Urban Villagers*, 337.

22. "Everyday Heroes."

23. The term *institutionally complete community* comes from Pyong Gap Min, "Koreans: An 'Institutionally Complete Community' in New York," in *New Immigrants in New York*, rev. ed., ed. Nancy Foner (Columbia University Press, 1985), 173–99.

24. Unfortunately, R.Dot took a lower profile around the time the group's founder actually moved to Battery Park City.

25. Gans, *Urban Villagers*.

26. For a fascinating comparison between fractions of classes that use education capital or financial capital as the basis of their distinctions, see Pierre Bourdieu, *Distinction: A Social Critique of the Judgement of Taste*, trans. Richard Nice (Cambridge, MA: Harvard University Press, 1984). See also discussion on the same topic by Herbert J. Gans in *Popular Culture and High Culture: An Analysis and Evaluation of Taste* (New York: Basic Books, 1999).

27. Gans, *Urban Villagers*, 332.

28. Ibid., 341.

29. News reports reinforced this understanding, which probably had some truth. "With Governor Pataki determined to break ground on July 4, work is moving ahead on the Freedom Tower." Robin Pogrebin, "The Incredible Shrinking Daniel Libeskind," *New York Times*, June 20, 2004.

30. In addition to *American Project*, see the film documentary on residents' displacement by Sudhir Venkatesh, *Dislocation* (2005).

31. Roy Rosenzweig and Elizabeth Blackmar, *The Park and the People: A History of Central Park* (Ithaca: Cornell University Press, 1998), 313.

32. Alexander Cooper Associates, *Battery Park City Draft Summary Report and 1979 Master Plan*, 1979, www.batteryparkcity.org/pdf_n/1979_Master_Plan .pdf, 65.

33. Ibid., 44.

34. Battery Park City Authority, *Annual Report*, 1984, Avery Library, Columbia University, 7.

35. Community involvement is described in Susan J. Fainstein, *The City Builders: Property, Politics, and Planning in London and New York* (Cambridge, MA: Blackwell, 1994), 169. Activities are listed in Battery Park City Authority, *Annual Report*, 1989–90, 6.

36. Resolution approved at a joint meeting of the Battery Park City, Financial District, and World Trade Center Redevelopment Subcommittees of Community Board 1, July 1, 2004.

37. William H. Whyte, *The Social Life of Small Urban Spaces* (Direct Cinema Limited), 1988.

38. Ida Susser, *Norman Street: Poverty and Politics in an Urban Neighborhood* (New York: Oxford University Press, 1982), 203.

Notes to Chapter 7: Definitely in My Backyard

1. Richard Sennett, *The Conscience of the Eye: The Design and Social Life of Cities* (New York: W. W. Norton, 1990); Ida Susser, *Norman Street: Poverty and Politics in an Urban Neighborhood* (New York: Oxford University Press, 1982), 193.

2. Francis Russell, "Battery Park City: An American Dream of Urbanism," in *Design Review: Challenging Urban Aesthetic Control*, ed. Brenda Case Scheer and Wolfgang F. E. Preiser (New York: Chapman and Hall, 1994), 208.

3. A long tunnel would have kept through traffic headed to and from the Brooklyn-Battery Tunnel off surface streets. But one main reason offered by DOT planners for not building the long tunnel was the paradoxical finding that the longer a tunnel is, the fewer motorists can use it. Most of the traffic on this part of West Street, by their calculations, got on or off somewhere along the tunnel route, and therefore would have taken surface streets rather than the tunnel.

4. At least four polls of Downtown residents included questions about the tunnel. The most comprehensive is J. Trichter and C. Paige, "The Rebuilding of Lower Manhattan: As Plans Progress, Lower Manhattan Residents Evaluate," Pace Poll Report, Pace University, March 15, 2004. Others are Community Board 1, "CB1.org Survey," 2003, www.cb1.org; and Friends of Community Board 1, "Downtown Residential Poll," 2004. I was also able to analyze the unpublished

data from a poll conducted online in 2003 by the community group Battery Park City United. Though the methods and goals of the polls differed, analysis showed strong opposition to the tunnel and greater opposition in Battery Park City than in neighboring areas.

5. "Pollsters Dispute Express' West St. Survey Analysis," *Downtown Express*, June 17–23, 2003.

6. Battery Park City United, unpublished data from online poll, 2003 (see n. 4 above); Community Board 1, "CB1.org Survey"; Friends of CB1, "Downtown Residential Poll,"; Trichter and Paige, "Rebuilding of Lower Manhattan."

7. Though comments on this bulletin board were anonymous, I take them as genuine posts by community members for several reasons. First, printouts of them were given to me by a member of the Coalition to Save West Street as accurate records of discussions by residents. Second, the IP addresses recorded with each posting concur with those I have recorded for some of the Battery Park City residents I know. Third, in interviews residents repeatedly referred to this site as one they used, and throughout the extended online exchanges no one questioned the identity of participants who claimed to be Battery Park City residents. (Participants on other sites had questioned the accuracy of other claimed identities.) Participants also provided biographical background (street addresses, locations of workplaces, and other details) that were consistent with those of known residents and tunnel activists.

8. Whatever the past policies at the Winter Garden Mall, it would have been unlikely that anyone would have been able to sleep there overnight in the period of my study, unless very well hidden. Mall personnel did not equivocate when they said they removed apparently homeless people from the mall.

9. Community Board 1 Manhattan, Resolution on West Street Short By-Pass, www.nyc.gov/html/mancb1/downloads/pdf/Resolutions/03-07-29.pdf, July 29, 2003.

10. One example of a group losing its battle is that of the businesses occupying Downtown's Radio Row that failed forty years earlier to stop the Port Authority from condemning their buildings to make room for the World Trade Center. James Glanz and Eric Lipton, *City in the Sky: The Rise and Fall of the World Trade Center* (New York: Times Books, 2003), 171–73.

11. David W. Dunlap, "Aspirations Bump into Practicalities at Ground Zero," *New York Times*, March 6, 2003, B3. One unnamed Lower Manhattan Development Corporation official said of the competing master plans, "Fundamentally it's a sideshow because none of these things will be built." In Charles Bagli, "Architects' Proposals May Be Bold, but They Probably Won't Be Built," *New York Times*, December 19, 2002, B11.

12. Coalition to Save West Street, "Governor Pataki and Batter Park City Authority President Carey Want a Giant Tour Bus Depot in the Middle of Battery Park City," flyer, fall 2003.

13. "LMDC Nibbles at BPCA: Now Officially Part of the WTC Plan, Site 26 May Become a Bus Depot. Residents Protest, Envisioning Lines of Idling Buses in Battery Park City," *Battery Park City Broadsheet*, September 25–October 10, 2003, 1.

14. I use *NIMBY* with hesitation, given Gibson's critique of the popular conception of it. But even if the term is freed from the stigma attached to it, I expect the difference remains between the relative power of those who can say "no" to a project and those who can say "where." T. Gibson, "NIMBY and the Civic Good," *City and Community* 4, no. 4 (2005): 381.

15. Douglas S. Massey and Nancy A. Denton, *American Apartheid: Segregation and the Making of the Underclass* (Cambridge, MA: Harvard University Press, 1993), 76; Jane Jacobs, *The Death and Life of Great American Cities* (New York: Vintage, 1961).

Notes to Chapter 8: Conclusion

The chapter's epigraph is taken from Ada Louise Huxtable, "Who's Afraid of the Big Bad Buildings?" *New York Times*, May 29, 1966, D13.

1. "Top Projects Started," New York Construction, June 2006, www.newyork .construction.com/projects/TopPrj_05-06/TPstrt1-5.pdf.

2. The fashion to require parks "pay for themselves" has been enshrined in law, at least since Congresswoman Nancy Pelosi (D-CA) conceded to congressional Republicans and argued that although the U.S. government had invested in San Francisco's Presidio for 150 years while it was a military base, to become a park it should suddenly "pay for itself." Amy Meyer and Randolph Delehanty, *New Guardians for the Golden Gate: How America Got a Great National Park* (Berkeley: University of California Press, 2006), 263. The political expedience of expecting parks to pay for themselves has led the trend to spread from coast to coast. The 1998 legislation creating New York City's Hudson River Park required that "to the extent practicable" commercial operations in the park cover operation and maintenance costs. New York State Legislature, "Hudson River Park Act," 1998 Sess. NY Legis Ch. 592 (S. 7845) (McKinney's 1998 Session Law News of New York, 1998), www.hudsonriverpark.org/pdfs/act/act.pdf.

3. Cambridge Systematics, Inc., "The Highway Construction Equity Gap," report for Texas Department of Transportation Government and Public Affairs Division, February 2008, in possession of the author.

4. Such high gas tax rates would allow gas taxes to cover the cost of roads only if consumers continued to buy the same quantity of gas, at those higher prices, as they do today. Assuming drivers would use less high-priced gas, there may be no way to pay for a road except via subsidies. This also appears to be the case, in this study, for high-quality public space. See Owen D. Gutfreund, "Driving Takes Its Toll," *New York Times*, September 4, 2004.

5. Michael B. Katz, *The Undeserving Poor: From the War on Poverty to the War on Welfare* (New York: Pantheon, 1989).

6. New York City Department of Housing, Preservation and Development, "Apartment Seekers: Mitchell Lama Housing," n.d., www.nyc.gov/html/hpd/html/apartment/mitchell-lama.shtml, accessed October 21, 2010.

7. For examples of coded racial language, see Maria Kefalas, *Working Class Heroes: Protecting Home, Community, and Nation in a Chicago Neighborhood* (Berkeley: University of California Press), 27, 41.

8. On school boundaries, see Kevin Fox Gotham, "Beyond Invasion and Succession: School Segregation, Real Estate Blockbusting, and the Political Economy of Neighborhood Racial Transition," *City and Community* 1, no. 1 (2002): 83–111.

9. The term is Pierre Bourdieu's, from *The Logic of Practice* (Palo Alto: Stanford University Press, 1992), 133.

10. U.S. Census, American Community Survey 2005–9, SocialExplorer.com, 2010.

11. Mark K. Levitan and Susan S. Wieler, "Poverty in New York City, 1969–99: The Influence of Demographic Change, Income Growth, and Income Inequality," *FRBNY Economic Policy Review*, July 2008, www.ny.frb.org/research/epr/.

12. David L. Gladstone and Susan S. Fainstein, "The New York and Los Angeles Economies," in *New York and Los Angeles: Politics, Society, and Culture, A Comparative View*, ed. David Halle (Chicago: University of Chicago Press, 2003).

13. Levitan and Wieler, "Poverty in New York City."

14. Theresa Devine, "New York City's Long-Term Unemployment Rate Continues to Outpace U.S. Rate," Independent Budget Office, IBO Web Blog, July 30, 2010, http://ibo.nyc.ny.us.

15. "The Jobless Rate for People Like You," *New York Times*, November 6, 2009.

16. At times when the New York and U.S. unemployment rates have been comparable, it has been because both rates have shot up precipitously but merge at a high level. That has been the case with the sharp increase since 2008. See graph accompanying Devine, "New York City's Long Term Unemployment Rate."

17. Mark Levitan, "It Did Happen Here: The Rise in Working Poverty in New York," in Halle, *New York and Los Angeles*, 254.

18. Roland Li, "Is the Rent Too Damn High?" *Real Estate Weekly*, November 1, 2010.

19. Al Amateau, "Wildman Wonders: Wildman and Parks Reach an Accord," *Chelsea Clinton News*, May 1, 1986: "Three weeks before, the 37-year-old expert, enthusiast and advocate of wild edible plants was arrested, handcuffed and booked for the criminal misdemeanor of picking and eating weeds in the park. . . . But Wildman's tour on Sunday was sanctioned by the Parks Department."

20. By democratic processes, I refer not just to voting but to politicians' need to be (somewhat) responsive to (some large number of) constituents, and the ability of a broad public to participate politically.

21. For examples of nostalgia for Moses, see media coverage of Hilary Ballon and Kenneth T. Jackson, eds., *Robert Moses and the Modern City: The Transformation of New York* (New York: W. W. Norton, 2007).

22. Benjamin Barber, *Strong Democracy: Participatory Politics for a New Age* (1984; repr., Berkeley: University of California Press, 2003), 141.

23. It is relevant to point out that the public spaces Moses is most known for, such as pools and playgrounds, were single-use spaces. Often they were a use that fit into a larger park that already served multiple purposes. Without rich community input, Moses would have had difficulty compiling the information needed to create a complex space that would match the activity needs of its surrounding community.

24. One of the inherent problems of "devolution" is that small-town governance is often less democratic than large-scale processes because a local board excludes everyone outside that geographic area, even when their decisions will affect outsiders.

Notes to Appendix B

1. Battery Park City United, unpublished data from online poll, 2003 (see ch. 7, n. 4 above); Community Board 1, "Downtown Residential Poll," 2004; Friends of Community Board 1, "CB1.org Survey," 2003; J. Trichter and C. Paige, "The Rebuilding of Lower Manhattan: As Plans Progress, Lower Manhattan Residents Evaluate," Pace Poll Report, Pace University, March 15, 2004.

2. Frances Fox Piven and Richard Cloward, *Poor People's Movements: Why They Succeed, How They Fail* (New York: Vintage, 1978); Verta Taylor, "Social Movement Continuity: The Women's Movement in Abeyance," *American Sociological Review* 54 (1989): 761–75.

3. Mitchell Duneier, *Sidewalk* (Farrar, Straus and Giroux, 1999), 348.

4. William Foote Whyte, *Street Corner Society: The Social Structure of an Italian Slum*, 4th ed. (Chicago: University of Chicago, 1993, 1943). An interesting example of pseudonyms: both Whyte and Elijah Anderson call the city they are researching "Eastern City," yet Whyte's is recognizably Boston just as Anderson's is clearly Philadelphia.

5. Duneier, *Sidewalk*, 348.

Index